Treatment System Hydraulics

Other Titles of Interest

Advances in Water and Wastewater Treatment, **edited by Rao K. Surampalli and K. D. Tyagi** (ASCE Committee Report). State-of-the-art information on the application of innovative technologies for water and wastewater treatment with an emphasis on the scientific principles for pollutant or pathogen removal. (ISBN 978-0-7844-0741-7)

Biological Nutrient Removal (BNR) Operation in Wastewater Treatment Plants, **by ASCE and WEF** (ASCE Manual). Instructs readers in the theory, equipment, and practical techniques needed to optimize BNR in varied environments. (ISBN 978-0-0714-6415-4)

Design of Water Resources Systems, **by Patrick Purcell** (Thomas Telford, Ltd.). Comprehensive coverage of the assessment, development, and management of water resources engineering infrastructure. (ISBN 978-0-7277-3098-5)

GIS Tools for Water, Wastewater, and Stormwater Systems, **by Uzair Shamsi** (ASCE Press). Guidelines to developing GIS applications for water, wastewater, and stormwater systems. (ISBN 978-0-7844-0573-4)

Gravity Sanitary Sewer Design and Construction, Second Edition, **edited by Paul Bizier** (ASCE Manual). Theoretical and practical guidelines for the design and construction of gravity sanitary sewers. (ISBN 978-0-7844-0900-8)

Water Resources Engineering: Handbook of Essential Methods and Design, **by Anand Prakash** (ASCE Press). Practical methods to solve problems commonly encountered by practicing water resources engineers in day-to-day work. (ISBN 978-0-7844-0674-8)

Treatment System Hydraulics

John Bergendahl, Ph.D., P.E.

Reston, VA

Library of Congress Cataloging-in-Publication Data

Bergendahl, John.
  Treatment system hydraulics / John Bergendahl.
    p. cm.
  Includes bibliographical references and index.
  ISBN-13: 978-0-7844-0919-0
  ISBN-10: 0-7844-0919-6
1. Pipelines. 2. Hydraulic engineering. 3. Fluid dynamics.
4. Sewage disposal plants—Design and construction. 5. Water
treatment plants—Design and construction. 6. Water—Purification.
7. Sewage—Purification. I. Title.

  TC174.B46 2008
  628.1'4—dc22

                                             2008006299

Published by American Society of Civil Engineers
1801 Alexander Bell Drive
Reston, Virginia 20191
www.pubs.asce.org

Solutions to the end-of-chapter problems are available to instructors. Requests should include your office mailing address and be submitted to Publications Marketing, ASCE (address above) or marketing@asce.org.

Contents

Preface

Treatment systems may consist of many physical, chemical, and biological processes coupled together to achieve some overall treatment goal. These systems may be designed and operated for treating water for potable use, treating domestic and industrial wastewater prior to discharge, treating water for water reuse, purifying water for industrial purposes, etc. There are many textbooks and courses that cover the fundamentals of the physical, chemical, and biological processes that make up these treatment systems. Yet often the most challenging design, construction, and operational problems in treatment systems are due to the hydraulics of the system. Will the pipe or channel achieve design flow? What is the proper valve to use for a certain application? How are pumps chosen? How is the system behavior controlled? What are the proper materials to use? Engineers involved with treatment systems have to know how to answer these types of hydraulics questions and many more.

Although there are a plethora of courses and textbooks that cover general fluid mechanics and general hydraulics, there is very little instruction at most engineering schools on hydraulics specifically for treatment systems. This text was created for the author's course at Worcester Polytechnic Institute on treatment system hydraulics when a suitable text could not be found that covered the salient hydraulics issues for treatment systems. This text covers the "nuts and bolts" of treatment systems, which is what most entry-level engineers and many experienced engineering practitioners deal with on a day-to-day basis.

This text has chapters on the topics that should be of great utility for engineers in addressing hydraulics of treatment systems. Chapter 1 presents an introduction to treatment systems and hydraulics as background material. The material in Chapter 1 may already be familiar to those either with more experience in treatment process course work or with experience as an engineering practitioner in the field. Chapters 2, 3, and 4 cover material that is fundamental to subsequent chapters and is needed for understanding hydraulics design and troubleshooting. Chapter 2 is on fluid properties, Chapter 3 reviews fluid statics, and Chapter 4 covers fundamentals of fluid flow. A significant part of the text that is of great importance to treatment system engineers is in Chapter 5 on friction in closed-conduit fluid flow, Chapter 6 on pumps and motors, Chapter 7 on fluid flow in granular media, and Chapter 8 on valves. Instrumentation provides much operational information for engineers and operators to troubleshoot their systems and is discussed in Chapter 9. Engineers must specify materials for treatment systems with knowledge of the potential corrosion conditions expected in the system, and how materials will perform in those conditions. Details of piping materials and corrosion are covered in Chapter 10. Treatment systems may not operate at steady-state conditions continuously. There may be increased system demands, pumps cycling on and off, valves opening and closing, and other scenarios where transients may occur in treatment systems. The subject of fluid transients is introduced in Chapter 11. And the final chapter of the text, Chapter 12, covers open channel flow because open channels are frequently used to convey waters in treatment systems.

Both the U.S. Customary System and the International System of Units (SI) are used by engineers and scientists around the world. Engineering practitioners in the United States have historically used the U.S. Customary System and the American engineering industry has resisted adoption of SI units. Yet the rest of the world uses the International System. With increasing collaboration among engineers on a global basis, American-based engineers may be involved in projects around the world requiring the use of the SI units. It is therefore imperative that engineers today be able to work in both systems. This text employs both the SI and U.S. Customary System units. Engineers today must be "ambidextrous" with respect to units and should be able to use both systems with little thought or confusion.

I thank the many students in my Treatment System Hydraulics course at WPI who were exposed to the draft version of this text, and who provided many constructive comments. The feedback certainly resulted in a much better text. I also thank my family, especially my wife Kimberly, who provided great support and inspiration throughout this effort.

CHAPTER 1

Introduction to Treatment Systems and Hydraulics

Chapter Objectives

1. Describe the various types of treatment systems in use.
2. Identify the importance of hydraulics in treatment systems.

Treatment systems are used in many applications around the world. Systems may be engineered for treating solid phases (such as contaminated soil), air phases (such as emissions from power plants), and liquid phases (such as contaminated water). The focus of this text is on the hydraulics of systems designed and constructed to treat liquids, primarily aqueous phase systems. Aqueous-phase, engineered treatment systems are employed to modify water quality before delivery as a commodity (i.e., potable water) or after a use that adds undesirable constituents to the water (i.e., domestic wastewater). This modification in water quality is brought about by a series of individual actions or steps in a treatment system called processes. Processes employed in treatment plants can be physical, such as sedimentation or filtration, biological, such as trickling filters or activated sludge, or chemical, such as disinfection or pH adjustment. The processes that are applied in treatment systems depend on the water quality changes desired, and technically effective and economically efficient processes and systems are preferred.

Engineered systems (see Fig. 1-1) are needed to do the following:

- treat water extracted from a water source (fresh water or sea water) to produce drinking water,
- remove pathogens, organics, and nutrients from domestic wastewater before discharge,
- pretreat industrial effluents before discharge to domestic wastewater treatment plants,
- highly purify water for specialized purposes (e.g., for use in nuclear and pharmaceutical industries),
- remove hazardous compounds (e.g., organics, metals, and radionuclides) from pump-and-treat groundwater remediation projects, and
- treat water for direct or indirect reuse for various purposes (e.g., agriculture, flushing water closets, and cooling).

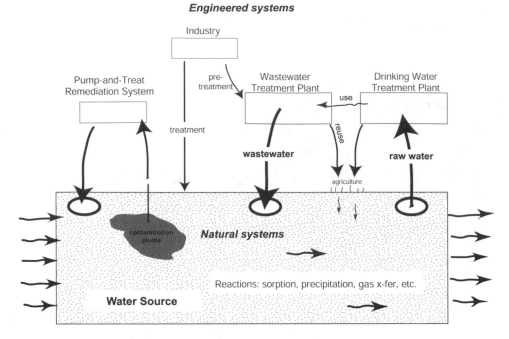

Figure 1-1. Engineered and natural water systems.

In these systems, water is neither created nor destroyed, but the undesirable constituents are changed in form or destroyed with physical, chemical, and/or biological processes. A system for the removal of solids from drinking water with ultrafiltration, a physical process, is shown in Fig. 1-2. Figure 1-3 illustrates a chlorine contact tank, a chemical process used in a wastewater treatment plant. The water slowly flows through the open channels of the contact tank, providing time for the chlorine to contact the remaining pathogens in the treated water. Activated sludge is a biological process commonly employed in domestic wastewater treatment plants. An aeration basin in an activated sludge system is shown in Fig. 1-4.

The processes that are typically used in treatment systems are classified into physical, chemical, and biological processes. In a physical operation, the application of physical forces modifies the water properties. Common physical processes include screening, flotation, gas stripping, gas absorption, filtration, sedimentation, and flocculation. Chemical processes employ chemical addition and/or reactions to bring about a water quality change. Chemical processes that are commonly used include chlorination, dechlorination, precipitation, advanced oxidation, and neutralization. Biological operations use biological activity for treatment. Common biological processes include activated sludge, trickling filters, and anaerobic digestion.

These individual physical, chemical, and biological processes are not used by themselves, but are usually "chained together" in series to bring about an overall change in water quality in the system. A schematic of a typical wastewater treatment system is illustrated in Fig. 1-5. First, the influent goes through primary sed-

Figure 1-2. Ultrafiltration system at a drinking water treatment facility.

Figure 1-3. Chlorine disinfection contact tanks in a wastewater treatment system. Chlorine is added immediately prior to the tanks, allowed to mix, and then the chlorine contacts the pathogens in the contact tank.

Figure 1-4. Activated sludge aeration tank in a domestic wastewater treatment facility.

imentation, where solids are removed by gravitational forces. Second, the wastewater passes into the aeration basin of the activated sludge process, after which it continues on into the secondary settling tank. In secondary settling, the biological mass created in the activated sludge aeration basin settles to the bottom of the tank. The wastewater may be subjected to advanced or tertiary treatment steps, such as nutrient removal. Finally, the wastewater is usually disinfected before discharge. The wastewater flows from step to step (process to process) through various conduits. Moreover, the sludge produced by the two sedimentation steps (primary and secondary) must be handled with fluid systems as well.

In a typical water treatment facility (shown in Fig. 1-6), water flows into a rapid mix tank where chemicals are added to destabilize the small particles in solution.

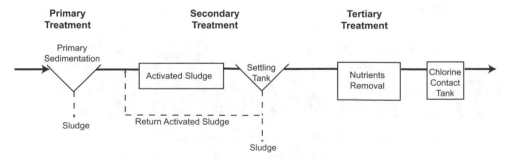

Figure 1-5. Schematic of a typical wastewater treatment facility.

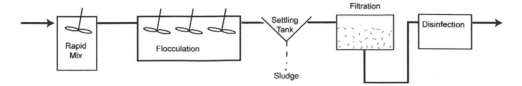

Figure 1-6. Schematic of a typical water treatment plant.

Next, the water passes into a flocculation tank, where collisions between the destabilized particles are promoted with slow mixing. The larger particles are then removed via sedimentation in a settling tank; this is followed by sand or mixed media filtration. The water is then disinfected (e.g., chlorinated) before distribution.

In all treatment systems, the water must travel from process to process in the system. The water may flow in open channels (see Fig. 1-7) or in pipes, induced by gravity or pressure. The effectiveness of the individual processes that are coupled together to make a system is very important, but predictable and efficient hydraulic behavior of the system is crucial. The water must flow from process to process at the proper flow rate, with no or minimal leakage from pipes, and without overflowing open channels. Its characteristics (pressure, flow rate, etc.) must be controllable (through, e.g., isolation of lines) and the flow must be well behaved. The science and application of hydraulic principles allow the practitioner to ensure that the system will operate successfully and efficiently.

The treatment system engineer must be concerned with the hydraulics of the water itself, as well as pumping chemicals to be added (e.g., high-strength acids and bases for pH adjustments), pumping sludge from sedimentation processes, etc. The nature of the fluid being transported affects the system hydraulic design.

Figure 1-7. Wastewater flowing through open channels to gates.

For example, sludge is a non-Newtonian fluid and this inherent property may be accounted for in the analysis of friction loss in a sludge pumping system. In modern treatment systems, hydraulics must be addressed for effective system performance and operation.

Systems such as these did not always exist. Unsanitary conditions occurred throughout the world without proper disposal of domestic and industrial wastes, and without purification of water before use. Wastes of all types ended up in the streets and open sewers, which were flushed with water as far back as Roman times (Babbitt 1932). Periodic flushing of the sewers was an attempt to keep them somewhat clean, although they were quite unsanitary, producing foul odors, insect breeding grounds, etc. The first large sewer system was constructed in New York City in 1805. Large sewer systems were also built in Paris starting in 1833 and Hamburg in 1842 (Babbitt 1932). Although the diversion of waste from streets to the sewer systems helped sanitation, treatment of the wastewater collected by the sewers was yet to come. Human waste was usually prohibited from the sewers until later, as it was considered unsafe. A pneumatic, vacuum collection system for domestic waste was employed in small, limited applications in European cities, but its implementation was not feasible on a wide scale (Babbitt 1932).

In England in the late 1800s, it was not uncommon for human waste to be present in "midden heaps" and for people to defecate in the road or in gardens (Corfield 1887). There just wasn't any place to dispose of refuse, excrement, ashes, and anything else that needed to be discarded. Eventually, open pits for human excrement, or cesspools, were constructed; frequently one cesspool served many homes. Unfortunately, these cesspools were often little more than ditches that ended up holding stagnant wastewater, and they frequently overflowed. Stone and brick lining assisted in the percolation of the liquid contents into the surrounding soil and provided some structural integrity. Different types of covering or roof for the cesspool kept rainwater from directly falling into the cesspools. Nonetheless, the contents often permeated the soil and foundations of nearby houses, and they often ended up contaminating drinking water wells. High incidences of typhoid fever, cholera, small pox, and other water-borne diseases were common (Corfield 1887). There is little doubt that the prevalence of water-borne diseases were in large part due to the methods in which domestic wastes were disposed of. Bear in mind the illustration of the water cycle shown in Fig. 1-1; that is, the water source may become contaminated by improper treatment and/or disposal of wastes. It was known even then that "the water pumped from the shallow wells in London is little else but filtered sewage" (Corfield 1887). However, when irrigation of agricultural crops with sewage was properly conducted through sewage farms, as had been done in England, France, and Germany in the 1800s (Corfield 1887), the water supply was protected, the sewage disposed of, and valuable crops were produced.

More advanced treatment processes were developed over time. Aeration of sewage was investigated by Dr. Angus Smith in 1882, Dr. Drown at the Lawrence Experiment Station (founded by the Massachusetts State Board of Health) in 1890, Colonel Waring at Newport, RI in 1892, and Clark and Gage at the Lawrence Experiment Station in 1912 (Babbitt 1932). The activated sludge process as we now know it was developed by Fowler, Ardern, and Lockett from England in 1913 after Fowler visited the Lawrence Experiment Station in 1912 (Babbitt 1932). The activated sludge process consists of two tanks: an aeration tank where the microorganisms that degrade organics in wastewater are "grown" and a sedimentation tank where they are removed through gravitational forces (settling). A full-scale activated sludge system was installed in Milwaukee in 1915 and in Houston in 1917. The activated sludge process has become ubiquitous in wastewater treatment systems owing to its effective treatment of wastewater. However, it requires the collection of sludge from the sedimentation tank, and the pumping of the sludge back to the aeration basin, a much more demanding and sophisticated hydraulic system than a simple batch or continuous flow biological process without sludge recycle. Of course, it also necessitates the provision of oxygen to the microorganisms in the aeration basin, with all the piping and valves needed for oxygen or air delivery to keep the biological consortia viable and consuming organics in the wastewater.

Increased sanitation and protection of water sources by construction of sewers to collect and divert wastewater and rainwater from streets produced advances in public health. For example, treatment of Thames River water supplied to London with sand filtration was shown to provide a 97.7% reduction in microorganisms in the supplied water (Corfield 1887). However, it was not known at the time (1886) that removal of pathogens was occurring with sand filtration. We now have a much better comprehension of the mechanisms of pathogen removal with sand filtration and an understanding of the performance of slow sand filtration such as in Fig. 1-8.

Figure 1-8. Slow sand filter used for drinking water treatment. The filter is under renovation with new sand being spread.

The physical, chemical, and biological processes are crucial, and it is very important for engineers to provide for proper hydraulic design of these systems.

Groundwater remediation schemes also make use of fluid systems to transport various fluids to and from contaminated areas in the ground. Acids, bases, and reactants may be delivered for in situ remediation systems, and contaminated water is pumped from contaminated aquifers to treatment systems in pump-and-treat systems. The treatment systems for these remediation plans may consist of various physical, chemical, and biological operations. Adsorption to granular activated carbon and air stripping are two common processes used in these systems. But again, systems like these must function effectively, process-wise and hydraulically.

Throughout the years, many have sought to understand the flow of water. This is certainly a practical quest, as the delivery of water to cities and areas for agriculture is an extremely important undertaking. It is also important to understand the flow of water in natural streams, rivers, and channels. The Persians laboriously hand dug underground aqueducts starting in 700 B.C.; these conveyed groundwater many miles from its source to cities and for agriculture (Wulff 1968). Large aqueducts, up to 90 km in length, brought water to baths and residences in Rome. The construction of these Roman aqueducts commenced in 312 B.C. and ended in A.D. 226 (Draffin 1939). Progress was also being made in other parts of the world on water supply and delivery. The Anasazi constructed dams and reservoirs in the southwest United States as early as A.D. 950. Water interceptor ditches captured water runoff, funneling it to reservoirs where the water could be used for domestic purposes or irrigation of crops. There is also evidence that the Anasazi used a settling basin to remove suspended material in an inlet just prior to the storage basin at Mummy Lake, in Mesa Verde National Park, Colorado (Breternitz 1999). The southwestern United States is mostly dry, and the collection and use of water by inhabitants was crucial for survival.

Inca engineers and tradesmen started construction of Machu Picchu in 1450 and very carefully created a water supply channel, fountains, and drain channels. The material of construction was stone, hand carved and assembled into channels critical for the survival of the population of the city. This network of water conveyance structures was well-preserved within the city with no maintenance for centuries after it was abandoned. Figure 1-9 shows a fountain that delivered water through an orifice outlet to a basin in stone. The orifice could supply a fluid velocity at the outlet so that the water would separate from the stone wall at a flow rate as low as 10 L/min (Wright and Zegarra 2000). In addition, the supply to some of the fountains was restricted with an orifice and bypass to prevent water flows that the fountains could not handle. The Inca designed the water supply to be segregated from waste, evidently aware of the potential for disease transmission.

Leonardo da Vinci, among his many other interests, delved into hydraulics at the end of the 1400s and the beginning of the 1500s. When studying the flow in a river, da Vinci noticed velocity distributions that were a function of the distance from a stationary wall: "Swifter becomes that part of the liquid which is farther from the friction against a denser body." Da Vinci designed pumps, siphons, water

Figure 1-9. Water fountain at Machu Picchu. *Source:* Wright and Zegarra (2000).

wheels, and many other hydraulic and nonhydraulic components. He envisioned the continuity principle: "In rivers, of breadth and depth whatever, it happens necessarily that, in any degree of length, the same amount of water passes in equal times" (Levi 1995). Da Vinci also correctly described the formation of vortices that form in arterial cavities (the sinuses of Valsalva) and cause the closure of heart valves. Unfortunately, da Vinci did not publish his works; it took many years for others to put his various notes and sketches together into the treatises that certainly contributed to the science of hydraulics as it is known today. Also, he did not collaborate with others in the field, preferring to work alone. So his thoughts remained unknown until long after his death. See Levi (1995) for more information on da Vinci.

Other early scientists and engineers contributed to the hydraulics body of knowledge. The pioneering work of Daniel Bernoulli, entitled *Hydrodynamics*, was published in 1738. Bernoulli was educated as a physician, yet he was deeply interested in fluids. His text is considered to be the beginning of fluid mechanics, and he discussed the "forces and motions of fluids." Bernoulli's father, Johann, was also fascinated with the motion of fluids and wrote *Hydraulics*, which was sent to Leon-hard Euler at the Academy of Sciences in St. Petersburg, Russia, in 1739. Both texts addressed the theory of fluid flow through an orifice, a problem previously unex-plained at the time. Johann Bernoulli's hydraulics theories were finally published in *Opera Omnia* (Complete Works of Johann Bernoulli) in 1743 (Levi 1995).

In the late 1700s, Antoine Chézy derived the well-known formula that bears his name for determining fluid velocity in an open channel as a function of slope, hydraulic radius, and a constant characteristic of the channel. In the late 1800s, Robert Manning refined this formula to specifically account for roughness of the channel walls. Interestingly, Manning thought the lack of unit consistency made his equation less desirable (Levi 1995), yet the Manning formula is still in wide-spread use today.

Jean Poiseuille, a physician in the 1800s, sought to understand the flow of blood in the human body. In 1842, he presented the landmark paper "Recherches expéri-mentales sur le mouvement des liquids dans les tubes de très petits diamèters" to the Academy of Sciences, providing experimental evidence that the velocity in round tubes is proportional to the diameter squared. The formula he derived to quantify flow through tubes is known as Poiseuille's law.

The contributions of Osborne Reynolds to our understanding of fluid flow can-not be overstated. In the late 1800s, Reynolds performed experiments to elucidate the occurrence of "sinuous" and "direct" flows in closed conduits. These flow char-acteristics are now known as "turbulent" and "laminar." The Reynolds number continues to be a crucial metric for understanding fluid flow characteristics and therefore frictional losses in pipes. Of course there are many other contributors to our body of knowledge on fluid statics, fluid mechanics, and hydraulics, including Archimedes, Euler, Galileo, Lagrange, Stokes, Torricelli, and Venturi (Levi 1995).

The treatment systems in use today were developed out of a strong need for clean water supply for domestic purposes and for improving the health of society and the environment. More sophisticated treatment processes developed over the years, and the hydraulic performance of these treatment systems, and collection and distribution systems, has become more important. Treatment systems must oper-ate effectively and predictably. In today's world of energy shortages, the hydraulic efficiency of fluid systems is also crucial to reduce energy consumption.

Problems

1. Describe the benefits that modern treatment systems provide to society.
2. Life expectancy has increased from about 28 years in classical Roman times, to 33 years in Medieval England, to 37 years in the late 19th century, to 50 years in the early 1900s, to approximately 80 years in Europe and the United States today. Discuss how dramatic increases in life expectancy can be attributed to treatment systems.
3. List as many activities that consume fresh water in your community as you can think of.

4. For many water uses, the water does not have to be as "clean" as water we may drink. Wastewater may be treated and reused in many freshwater applications not requiring drinking water standards. Identify opportunities for water reuse in your community.

5. List and describe some problems that could occur with a treatment system when it does not operate correctly from a hydraulic standpoint.

References

Babbitt, H. E. (1932). *Sewage and Sewage Treatment*, Wiley, New York.

Breternitz, D. A. (1999). *The 1969 Mummy Lake Excavations*, Wright Paleohydrological Institute, Boulder, CO.

Corfield, W. H. (1887). *The Treatment and Utilization of Sewage*, Macmillan, London.

Draffin, J. O. (1939). *The Story of Man's Quest for Water*, Garrard, Champaign, IL.

Levi, E. (1995). *The Science of Water. The Foundation of Modern Hydraulics*, ASCE Press, New York.

Wright, K. R., and Zegarra, A. V. (2000). *Machu Picchu. A Civil Engineering Marvel*, ASCE Press, Reston, VA.

Wulff, H. E. (1968). "The Qanats of Iran," *Sci. Am.*, 218(4), 94–105.

CHAPTER 2

Fluid Properties

Chapter Objectives

1. Identify fluid properties relevant to treatment system design and performance.
2. Predict shear stress for Newtonian fluids using Newton's law of viscosity.
3. Illustrate the behavior of non-Newtonian fluids.

In most treatment systems, fluids must be moved from one process or operation to another. The properties of fluids dictate the behavior of those fluids in treatment systems. This chapter covers the fundamental properties of fluids that affect their behavior and determine the hydraulic operational effectiveness of the system. The fluid properties that system designers are concerned with include density, the relationship between velocity gradient and shear stress, surface tension, and vapor pressure.

Fluids that we may encounter in treatment systems include the following:

1. aqueous-based liquids such as water, wastewater, sludge, acids, bases, and other treatment chemicals;
2. nonaqueous liquids such as gasoline, fuel oil, and lubricating oil; and
3. gases such as air, oxygen, and chlorine.

Although solid materials can withstand shear forces without permanent deformation, fluids may be continuously deformed by shear forces. When a fluid is subjected to an external shear force, the fluid will move until the force is removed. This is the characteristic behavior of a fluid that differentiates it from a solid—the fact that it may be continuously deformed by shear forces.

Density

The *density* of a fluid is an important property as it directly affects the weight of the fluid that is being handled. A fluid with a greater density weighs more than a fluid with a lower density of equal volume. The density can also affect the performance of systems in many other ways. The value of fluid density, which is a function of temperature and pressure, may be measured or looked up in known databases. See Fig. 2-1 for the density of water as a function of temperature. Density is defined as the fluid mass per unit volume that the fluid occupies:

$$\rho = \frac{m}{\mathbf{v}} \qquad (2\text{-}1)$$

Measuring the density of a fluid is done gravimetrically. A known volume of the fluid is weighed, the corresponding mass is discerned from the weight, and the density of the fluid is then calculated from Eq. 2-1. Density has units of mass/volume, such as kg/L, g/cm^3, kg/m^3, or lbm/ft^3.

With the exception of mercury, the liquids in Table 2-1 have a density three orders of magnitude greater than the gases. This greater density is the result of closer molecular spacing in the liquid phase. In addition, the change in density of gases with pressure variations is much greater than the change in density of liquids, as the molecular spacing can be affected by the pressure; gases can be compressed. In fact, liquids are usually assumed to be incompressible because the molecular spacing, and therefore volume, is only slightly changed with pressure. A fluid is considered *incompressible* if the density change with pressure difference is insignificant. If the fluid density change is not insignificant, the fluid is considered *compressible*. Strictly speaking, all fluids are compressible to some degree, but we

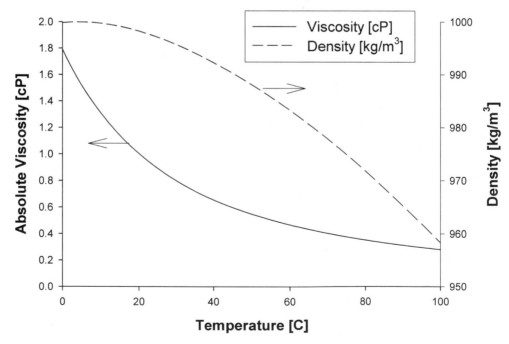

Figure 2-1. Density and absolute viscosity of water as a function of temperature. *Source:* Data from Lemmon et al. (2005).

Table 2-1. Density of selected fluids at 20 °C (68 °F) and 1 atm.

Fluid	Density, ρ [kg/m^3]
Water (l)	998.2
Benzene (l)	878.9
Ethylene glycol (l)	1,113
Glycerol (l)	1,261
Methanol (l)	791.0
Toluene (l)	866.8
Mercury (l)	13,550
Nitrogen (g)	1.165
Oxygen, O_2 (g)	1.331
Carbon dioxide, CO_2 (g)	1.839

Source: Data are taken from Lemmon et al. (2005), except for ethylene glycol and glycerol, which are from Speight (2005), and mercury, which is from Forsythe (2003).

make the incompressibility assumption for the liquids dealt with in treatment systems to simplify analyses.

A similar measure for molecular spacing often used as a surrogate for liquid density is the density of the liquid relative to the density of water, or the *specific gravity*. It is defined at a given temperature and pressure as

$$SG = \frac{\rho}{\rho_w (\text{at given } T \& P)} \qquad (2\text{-}2)$$

Whereas the density of a fluid has units associated with it, specific gravity has the advantage of being independent of the system of units being used. If the specific gravity of the liquid is known, the actual density is obtained by simply multiplying the specific gravity by the density of water at that temperature and pressure.

Relationship between Velocity Gradient and Shear Stress

The resistance of fluids to flow is characterized by the fluid viscosity. It should seem intuitive that water can be readily poured from one vessel to another. But thick ketchup may take some time to flow from its container, even with a good deal of "coaxing." It should be expected that the resistance to fluid flow will greatly affect the ability of fluids to flow from one treatment process to another. Many liquids that are encountered in treatment systems are considered "Newtonian"; that is, they follow Newton's law of viscosity.

Newton's Law of Viscosity

Consider a thin fluid film captured between two solid parallel plates, each with an area A, separated by a distance y_0, as shown in Fig. 2-2.

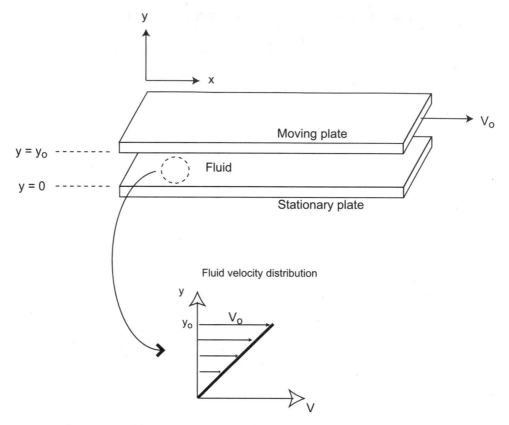

Figure 2-2. Sliding plates separated by a fluid.

The upper plate is moving at a constant velocity V_0. A constant force F on the upper plate is necessary to maintain V_0:

$$\frac{F}{A} = \mu \frac{V_0}{y_0} \tag{2-3}$$

where μ is a constant of proportionality, called the *viscosity*.

From this equation we can see that a lower force will result in a lower velocity V_0, with all other conditions the same, and a greater viscosity μ will produce a lower V_0 for a given applied force F. So a fluid with a greater viscosity will need a larger external force to achieve the same V_0. That indicates to us that fluids with greater viscosities will need greater forces, and therefore greater applied pressures, than those with low viscosities to get them to move to the same extent over time.

When the fluid velocity is low, the velocity profile of the fluid between the plates is linear:

$$\frac{dV}{dy} = \frac{V_0}{y_0} \tag{2-4}$$

According to Newton's law of viscosity, the shear stress on the fluid is then

$$\tau = \mu \frac{dV}{dy} \qquad (2\text{-}5)$$

So, for a Newtonian fluid, the shear stress is proportional to the velocity gradient dV/dy. The constant of proportionality, μ, is the *absolute viscosity* of the fluid, and it is independent of the velocity gradient. It is defined as

$$\mu = \frac{\tau}{\left(\dfrac{dV}{dy} \right)} \qquad (2\text{-}6)$$

Centipoise (cP) is the most commonly used unit for absolute viscosity, but poise (P) and Pa·s are also used; these are related by

$$100 \text{ centipoise (cP)} = 1 \text{ poise (P)} = 1\frac{\text{g}}{\text{cm} \cdot \text{s}} = 0.1 \, \text{Pa} \cdot \text{s} = 6.72 \times 10^{-2} \frac{\text{lbm}}{\text{ft} \cdot \text{s}}$$

The viscosities of some fluids at standard conditions are listed in Table 2-2.

The viscosity of gases is slightly dependent on pressure, and the pressure dependence can usually be ignored. However, the viscosity of a gas is strongly dependent on temperature (see Fig. 2-3), approximately following the relationship (Wilkes 1999)

$$\mu \cong \mu_0 \left(\frac{T}{T_0} \right)^n \qquad (2\text{-}7)$$

where μ_0 and n are parameters that depend on the gas (see Table 2-3).

Table 2-2. Viscosity of selected fluids at 20 °C and 1 atm.

Fluid	Viscosity, μ [cP]
Water (l)	1.0016
Benzene (l)	0.6816
Mercury, Hg (l)	1.55
Methane, CH_4 (g)	0.0110
Oxygen, O_2 (g)	0.0202

Source: Data are taken from Lemmon et al. (2005), except for mercury, which is from Bird et al. (1960).

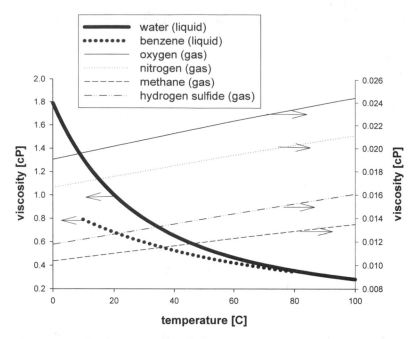

Figure 2-3. Absolute viscosities of representative liquids and gases as a function of temperature. *Source:* Data from Lemmon et al. (2005).

The viscosities of liquids are not affected by pressure but are dependent on temperature (see Fig. 2-3). Liquid viscosity approximately follows the relationship (Wilkes, 1999)

$$\ln(\mu) \cong a + b \ln(T)$$

or

$$\mu \cong e^{a+b \ln T} \tag{2-8}$$

Viscosity parameters for some common liquids are listed in Table 2-4.

Table 2-3. Viscosity parameters for gases.

Gas	μ_0 [cP]	n
Air	0.0171	0.768
Carbon dioxide	0.0137	0.935
Ethylene	0.0096	0.812
Hydrogen	0.0084	0.695
Methane	0.0120	0.873
Nitrogen	0.0166	0.756
Oxygen	0.0187	0.814

Source: Data are taken from Wilkes (1999).

Table 2-4. Viscosity parameters for liquids (with T in kelvins and μ in centipoise).

Liquid	a	b
Acetone	14.64	-2.77
Benzene	21.99	-3.95
Ethanol	31.63	-5.53
Glycerol	106.76	-17.60
Kerosene	33.41	-5.72
Methanol	22.18	-3.99
Octane	17.86	-3.25
Pentane	13.46	-2.62
Water	29.76	-5.24

Source: Data are taken from Wilkes (1999).

Another measure of the ratio of shear stress to velocity gradient is the *kinematic viscosity*. The kinematic viscosity of a fluid is the absolute viscosity divided by the fluid density:

$$v = \frac{\mu}{\rho} \tag{2-9}$$

The units used for kinematic viscosity are *centistokes* (cSt), stokes (St), m^2/s, and ft^2/s. These units are related by

$$1 \text{ centistoke (cSt)} = \frac{1 \text{ cP}}{1\dfrac{g}{cm^3}} = 10^{-6}\,\frac{m^2}{s} = 1.08 \times 10^{-5}\,\frac{ft^2}{s}$$

Non-Newtonian Fluids

Fluids that behave according to Newton's law of viscosity are called Newtonian fluids. However, the resultant velocity gradient is not proportional to the applied shear stress for all fluids; fluids for which Newton's law of viscosity does not hold are called *non-Newtonian fluids*. Examples of the types of non-Newtonian fluids, such as Bingham plastics, pseudoplastics, and dilatants, are shown in Fig. 2-4. The plots of shear stress versus velocity gradient for non-Newtonian fluids are not linear (and may have nonzero intercepts), whereas those of Newtonian fluids are linear with a zero intercept.

The non-Newtonian fluids are generally classified as Bingham plastics, pseudoplastics, and dilatants, and each of these fluids are described next.

Bingham Plastics

Bingham plastics are fluids that are not deformed at all by small shear stresses; they are only deformed by larger shears when the shear is above a threshold value.

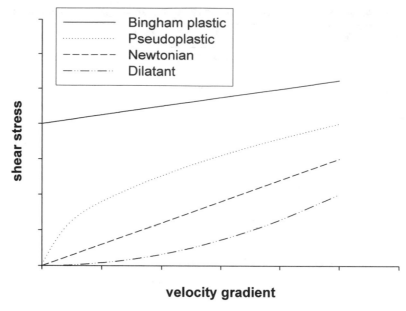

Figure 2-4. Behavior of Newtonian and non-Newtonian fluids under shear.

The fluid does not flow until the applied shear is above this given threshold. Examples of Bingham plastics are some slurries, jelly, and toothpaste. The shear behavior of Bingham plastics can be expressed with the Bingham model

$$\tau = \mu_0 \frac{dV}{dy} + \tau_0 \qquad (2\text{-}10)$$

where μ_0 is an "apparent" viscosity in the Bingham model and τ_0 is the threshold shear stress above which deformation occurs. These parameters are characteristics of the fluid and may be measured.

Pseudoplastics

Pseudoplastics are fluids that do not possess a constant apparent viscosity with respect to velocity gradient but exhibit shear stress that decreases from linearity with increasing velocity gradient. That is, the slope of the shear stress versus velocity gradient decreases with increasing shear rate, as the "apparent" viscosity is a function of shear rate. Pseudoplastics are represented with the concave downward curve shown in Fig. 2-4. Examples of pseudoplastics include some polymer solutions, blood, mud, and most slurries. The shear behavior of pseudoplastics can be described with the Ostwald–de Waele model, also called the *power-law model*:

$$\tau = \underbrace{m \left(\frac{dV}{dy} \right)^{n-1}}_{\text{"apparent viscosity"}} \frac{dV}{dy} \qquad (2\text{-}11)$$

Table 2-5. Ostwald–de Waele parameters for some pseudoplastic fluids.

Fluid	m $[lbf \cdot s^n \cdot ft^{-2}]$	n $[dimensionless]$
23.3% Illinois clay in water	0.116	0.229
33% lime in water	0.150	0.171
4% paper pulp in water	0.418	0.575

Source: Data are taken from Metzner (1956).

The constants m and n are parameters in the Ostwald–de Waele model and are properties of the specific fluid. The parameter n is dimensionless, and the units for m depend on the value of n. Table 2-5 lists some Ostwald–de Waele parameters for some fluids. For pseudoplastic fluids $n < 1$. For $n = 1$ and $m = \mu$, this equation simplifies to Newton's law of viscosity.

Wastewater sludge may be expected to behave as a pseudoplastic. The values for m and n depend on the type of sludge and the amount of solids present in the sludge, as well as other parameters (such as presence and amount of exocellular polymers). The parameters can be highly variable. Figures 2-5 and 2-6 illustrate the dependence of m and n on percent total solids for waste-activated sludge and anaerobically digested sludge. The figures also show the variability found in the values for the parameters.

Dilatants

Like pseudoplastics, dilatants are fluids that do not have a constant viscosity with respect to velocity gradient. The shear stress increases from that predicted by a linear relationship with increasing velocity gradient; the slope increases with

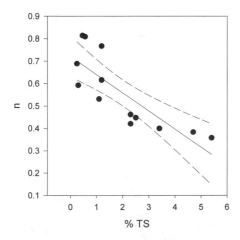
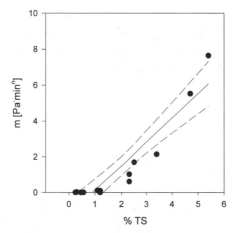

Figure 2-5. Power-law coefficients for waste-activated sludge as a function of percent solids. The solid line is from a linear regression and the dashed lines are 95% confidence intervals. *Source:* Data from Lotito et al. (1997).

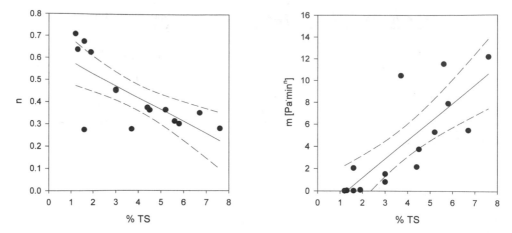

Figure 2-6. Power law coefficients for anaerobically digested sludge as a function of percent solids. The solid line is from a linear regression and the dashed lines are 95% confidence intervals. *Source:* Data from Lotito et al. (1997).

increasing shear rate. As for pseudoplastics, the Ostwald–de Waele (power-law) model can be used. For dilatant fluids $n > 1$. The values for m and n are constants that depend on the fluid. For more information on non-Newtonian fluids, see Bird et al. (1960).

Time Dependence of Viscosity

The viscosity of Newtonian fluids is time-independent; that is, the fluid viscosity does not change during the time that it undergoes deformation. Many non-Newtonian fluids also do not display shear stress versus velocity gradient responses that vary with time. However, some non-Newtonian fluids do have time-dependent properties. As illustrated in Fig. 2-7, the apparent viscosities (and shear stresses) of *thixotropic* fluids decrease with shearing time at constant velocity gradients, whereas the viscosities of *rheopectic* fluids increase with shearing time.

Surface Tension

At the interface between a gas phase and a liquid phase, the surface of the liquid tends to behave like an elastic membrane, or a "skin." Unequal attractive forces between molecules in the surface layer cause tension among the molecules, or *surface tension*, as depicted in Fig. 2-8.

Attractive forces between molecules, including van der Waals forces and hydrogen bonding (for compounds where hydrogen bonding can take place between molecules, such as water), are symmetric for a molecule within the bulk phase. But forces on a molecule at the interface between two phases, such as

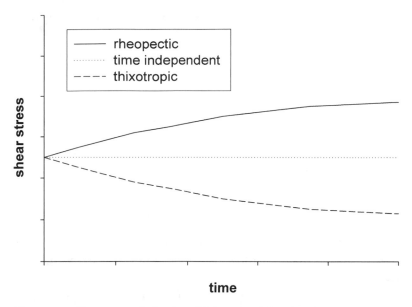

Figure 2-7. Shear stress behavior with increased shearing time at constant velocity gradients.

between the condensed (liquid) and gas phases, are not symmetrical. Van der Waals forces originate from spontaneous electrical and magnetic polarizations, which produce a fluctuating electromagnetic field between molecules and result in an electrodynamic interaction force that is usually attractive. The van der Waals interaction energy between two molecules is proportional to r^{-6}, the separation

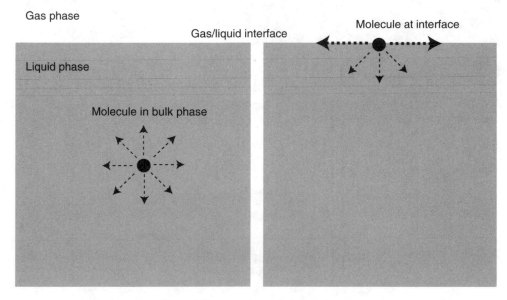

Figure 2-8. Symmetrical forces on a molecule in bulk phase and unsymmetrical forces on a molecule at the gas/liquid interface.

distance raised to the −6 power. So the interaction energy between two molecules drops off quickly with increasing separation distance. For the condensed phase (liquid) the molecules are close enough for the van der Waals force to be significant, whereas for the gas phase, the molecules are so far apart that the magnitude of the van der Waals interaction is insignificant in comparison to the magnitude in the liquid phase. So it should be expected that there will be significant asymmetry in the van der Waals forces between molecules at an interface between the gas and liquid phases.

Hydrogen bonding can also play a significant role in the attractive interaction between molecules in substances such as water. A water molecule has four hydrogen bonding sites—two proton donor sites (the two hydrogen atoms) and two proton acceptor sites from two lone electron pairs on the oxygen atom. See Fig. 2-9. Because of this unique arrangement of two proton donors and two proton acceptors, water has a tendency to form a tetrahedral-coordinated structure in the bulk phase. But when the bulk water phase is not continuous, such as at an interface between the bulk liquid and gas phases, this hydrogen-bonded structure is interrupted. This discontinuity in hydrogen bonding opportunities contributes to the asymmetric van der Waals forces to result in surface tension. However, recall that it is the presence of the interface that produces unequal forces (van der Waals and hydrogen bonding) on the molecules at the surface, which produces surface tension, as shown in Fig. 2-8.

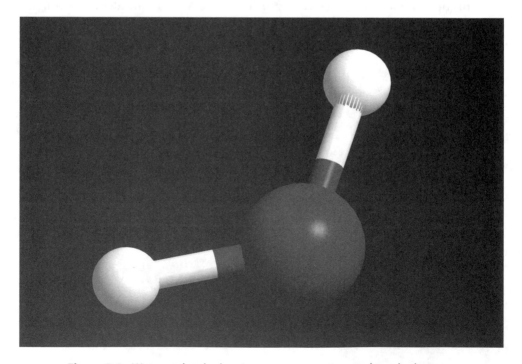

Figure 2-9. Water molecule showing one oxygen atom and two hydrogen atoms.

Table 2-6. Surface tension of selected fluids in air at 20 °C (68 °F).

Fluid	Surface tension [N/cm] (lbf/in.)
Acetic acid	2.8×10^{-4}
	(1.6×10^{-4})
Acetone	2.5×10^{-4}
	(1.4×10^{-4})
Benzene	2.8×10^{-4}
	(1.6×10^{-4})
Carbon tetrachloride	2.6×10^{-4}
	(1.5×10^{-4})
Ethyl alcohol	2.3×10^{-4}
	(1.3×10^{-4})
n-octane	2.1×10^{-4}
	(1.2×10^{-4})
Toluene	2.8×10^{-4}
	(1.6×10^{-4})
Water	7.5×10^{-4}
	(4.3×10^{-4})
Mercury	5.3×10^{-3}
	(3.0×10^{-3})

Source: Data are taken from de Nevers (1991).

This unsymmetrical overall force on the molecules creates a tension in the surface molecules that is manifested in a surface energy or surface tension. This surface tension can create a tendency for liquids to try to minimize their surface area and form a sphere.

The surface tension can be measured with many apparatus; one device is shown in Fig. 2-10. The surface tension of a film is the force exerted on the film, divided by the length over which the force is exerted:

$$\text{surface tension, } \sigma = \frac{\text{force}}{\text{length}} \tag{2-12}$$

We are concerned with surface tension because, in treatment systems, we may encounter two-phase flow where an interface exists between two phases (e.g., liquid and vapor, bubbles and drops), and this interface is characterized by a surface or interfacial tension.

Example

Water drops are discharged from a capillary tube that is 0.50 mm in diameter. See Fig. 2-11. What is the maximum drop volume expected upon drop detachment? Assume a temperature of 20 °C. This example is adapted from examples in Adamson and Gast (1997) and Middleman (1998).

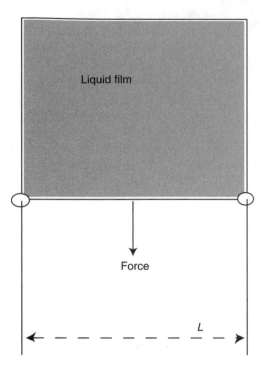

Figure 2-10. Measuring surface tension.

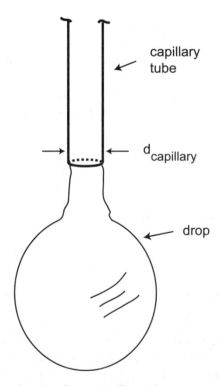

Figure 2-11. Drop hanging from a capillary tube.

Solution

When detachment occurs, the weight of the drop is just equal to the force that can be exerted by the surface tension at the neck of the drop (which has a diameter defined as $d_{capillary}$). In other words, the force from surface tension in the neck of the drop is what is holding the drop up. As the drop size gets larger and larger, eventually the surface tension can no longer hold the drop, and the drop falls. From Eq. 2-12, we have

$$\text{force from surface tension} = \sigma \cdot \text{length}$$
$$= \sigma \cdot (\pi \cdot d_{capillary})$$

The force from the mass of the drop (weight) is

$$\rho \cdot \mathbf{v} \cdot g$$

Setting the forces equal gives

$$\sigma \cdot (\pi \cdot d_{capillary}) = \rho \cdot \mathbf{v} \cdot g$$

Rearranging to solve for $\mathbf{v}$ we get

$$\mathbf{v} = \frac{\sigma \cdot (\pi \cdot d_{capillary})}{\rho \cdot g}$$

At 20 °C, $\sigma = 7.5 \times 10^{-4}$ N/cm, or 0.075 N/m, and $\rho = 998.21$ kg/m^3. Substituting in numerical values yields

$$\mathbf{v} = \frac{0.075\frac{N}{m} \cdot (\pi \cdot 0.50 \times 10^{-3}\,m)}{998.21\frac{kg}{m^3} \cdot 9.81\frac{m}{s^2}} \left[\frac{1\frac{kg \cdot m}{s^2}}{1\ N} \right] = 1.20 \cdot 10^{-8}\ m^3 = 12\ mL$$

It has been found that this strictly theoretical approach overpredicts the drop volume. In reality, a portion of the drop remains attached at the capillary tip and does not detach with the drop. It has been found that up to 40% of the drop volume may not detach with each drop (Adamson and Gast 1997).

Vapor Pressure

Treatment system engineers may be concerned with the presence of liquid and gas phases simultaneously as well as the transformation from one phase to another.

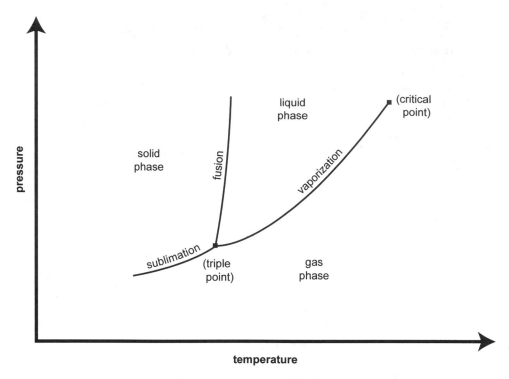

Figure 2-12. *PT* diagram for a general substance.

Figure 2-12 shows the behavior of a general pure substance as a function of temperature and pressure. The vaporization curve gives the liquid–gas equilibrium relationship. The line describes the temperature and pressure at which the liquid and gas phases coexist. If the pressure of a liquid at a given temperature (above the triple point where solid, liquid, and gas coexist) is reduced to that on the vaporization curve, the liquid will vaporize. So at a given temperature, there is a pressure, the *vapor pressure*, P_{sat}, where a liquid will be vaporized, and a gas will be condensed to liquid. The pressure versus temperature (*PT*) diagram for liquid and vapor phases of water above the triple point is shown in Fig. 2-13.

Equations to relate vapor pressure to temperature include the Clapeyron equation,

$$\log P_{sat} = A - \frac{B}{T} \tag{2-13}$$

where *A* and *B* are empirical constants that depend on the substance, the Antoine equation,

$$\log P_{sat} = A - \frac{B}{T+C} \tag{2-14}$$

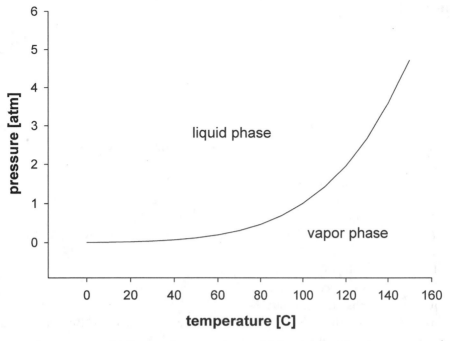

Figure 2-13. *PT* diagram for water above 0 °C showing liquid and vapor phases. *Source:* Data from Lemmon et al. (2005).

where A, B, and C are empirical constants dependent on the substance (see Table 2-7 for parameter values for some selected substances), and the Riedel equation,

$$\log P_{\text{sat}} = A - \frac{B}{T} + D \ln T + FT^6 \qquad (2\text{-}15)$$

where A, B, D, and F are empirical constants.

Table 2-7. Antoine equation parameters (for P in bars and T in kelvins).

Substance	A	B	C	Temperature range [K]
Acetone	4.42448	1312.253	−32.445	259–508
Acetic acid	4.68206	1642.540	−39.764	290–391
Ethanol	5.37229	1670.409	−40.191	273–352
Water	5.40221	1838.675	−31.737	273–303
Water	5.20389	1733.926	−39.485	304–333

Source: Data are taken from Linstrom and Mallard (2001).

Symbol List

a, b	empirical parameters
A	area
A, B, C, D	empirical constants in Clapeyron, Antoine, and Riedel equations
F	force
m	mass
n	empirical exponent
m, n	non-Newtonian fluid power-law coefficients
P	pressure
P_{sat}	vapor pressure
SG	specific gravity
T	absolute temperature
T_0	absolute reference temperature; $T_0 = 273.15$ K $= 459.67$ °R
$\mathbf{v}$	volume
V, V_0	velocity
y, y_0	distance
μ	absolute viscosity
μ_0	absolute viscosity at T_0; "apparent" viscosity in Bingham model
ν	kinematic viscosity
ρ	density
ρ_w	density of water
σ	surface tension
τ	shear stress
τ_0	threshold shear stress in Bingham model

Problems

1. Why should we be concerned with vapor pressure when we are designing a fluid system?
2. Water is in equilibrium at a pressure of 0.025 at 10 °C. What phase, liquid or gas, is it in? What phase would it be in at 30 °C? Justify your answers.
3. Using Eq. 2-8, calculate the viscosity of benzene in centipoise at 20 °C.
4. Waste-activated sludge consisting of 4% solids is subjected to a constant velocity gradient of 25 s^{-1}. What are the shear stress and apparent viscosity?

References

Adamson, A. W., and Gast, A. P. (1997). *Physical Chemistry of Surfaces*, Wiley, New York.

Bird, R. B., Stewart, W. E., and Lightfoot, E. N. (1960). *Transport Phenomena*, Wiley, New York.

de Nevers, N. (1991). *Fluid Mechanics for Chemical Engineers*, McGraw-Hill, New York.

Forsythe, W. E. (2003). *Smithsonian Physical Tables*, Knovel, Norwich, NY.

Lemmon, E. W., McLinden, M. O., and Friend, D. G. (2005). "Thermophysical Properties of Fluid Systems," *NIST Chemistry WebBook, NIST Standard Reference Database Number 69*, P. J. Linstrom and W. G. Mallard, Gaithersburg, MD, National Institute of Standards and Technology (http://webbook.nist.gov).

Linstrom, P. J., and Mallard, W. G. (2001). *NIST Chemistry WebBook, NIST Standard Reference Database Number 69*, National Institute of Standards and Technology, Gaithersburg, MD.

Lotito, V., Spinosa, L., Mininni, G., and Antonacci, R. (1997). "The Rheology of Sewage Sludge at Different Steps of Treatment," *Water Sci. Technol.*, 36(11), 79–85.

Metzner, A. B. (1956). *Advances in Chemical Engineering*, Academic Press, New York, 103.

Middleman, S. (1998). *An Introduction to Fluid Dynamics*, Wiley, New York.

Speight, J. G., ed. (2005). *Lange's Handbook of Chemistry*, McGraw-Hill, New York.

Wilkes, J. O. (1999). *Fluid Mechanics for Chemical Engineers*, Prentice Hall, Upper Saddle River, NJ.

CHAPTER 3

Fluid Statics

Chapter Objectives

1. Identify relevant pressure datums.
2. Calculate the pressure dependence from fluid depth in static fluids.
3. Quantify forces on surfaces resulting from fluid pressure.

A column of water will have a greater pressure with increasing water depth. For example, swimmers feel greater pressure in their ears as they dive deeper in a pool, and a submarine is exposed to higher pressures external to its hull as it travels deeper in the ocean. Dams (Fig. 3-1) will have high fluid pressures exerted at the base of the dam owing to the increasing pressure from the water depth on the upstream side of the dam. Fluid statics deals with the variation of pressures in piping systems, containers, impoundments, etc., when the fluid is not moving. Quantifying the pressure as a function of fluid depth is necessary for many reasons. These pressures must be considered by engineers so that material will not rupture and so that the energy inherent in the pressure is accounted for and/or properly used. We will cover many applications later in the text where the static pressure must be known.

Water distribution systems frequently have water towers and storage tanks, such as in Fig. 3-2, incorporated in the systems to (a) store water for emergencies and (b) provide pressure for high-demand situations. Water towers can enable the water distribution system to function for a time during drought conditions and when water is needed for fighting fires.

Fluid Pressure

The pressure exerted on a structure by fluids is an important concept for engineers to understand as the structure may rupture if the forces from the fluid pressure are too high. Consider the water tanks in Fig. 3-2. If the forces on the structural elements of the tanks (the curved plate comprising the tank sections and the rivets or bolts holding the sections together) are greater than the material can withstand,

Figure 3-1. The New Croton Dam. *Source:* Aqueduct Commission (1907).

the material may yield and rupture, causing a catastrophe. The normal force F_n resulting from a pressure difference between the inside and outside of a flat surface is related to the area A over which the pressure P acts according to

$$F_n = P \cdot A \qquad (3\text{-}1)$$

The normal force can be significant even with moderate pressures if the area A is large.

Example

Consider a pipe with a diameter of 6 in. (15.2 cm) with a closed end such as in Fig. 3-3. An internal pressure of 750 psi (or 51 atm, a typical pressure for feed to reverse osmosis membranes used in desalination) can produce a normal force on the end cap of

$$F_n = 750\,\frac{\text{lbf}}{\text{in.}^2} \cdot \left(\frac{\pi \cdot (6\ \text{in.})^2}{4} \right) = 21{,}000\ \text{lbf (or } 9.3 \times 10^4\ \text{N)}$$

This force must be withstood by the material of the pipe and the end connection to avoid failure. If the end cap is welded on, the welds between the end cap

(a)

(b)

Figure 3-2. Water storage tanks.

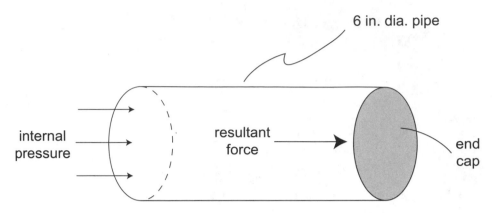

Figure 3-3. Internal pressure on a pipe end cap.

and the pipe must withstand the stress from the applied force. If the end cap is threaded on the pipe, the threads must withstand the force.

In considering the ramifications of pressure at an arbitrary point within a water column, the *direction* that the pressure acts must be preliminarily addressed to discern its importance. Consider a fluid of a given mass and volume that is not moving; that is, it is *static*. Pressure exists at all points in the fluid element that is occupied by the fluid shown in Fig. 3-4. The fluid element is a wedge-shaped volume that is dz thick.

The fluid element has gravitational and pressure forces acting on it, and the sum of the external forces must be zero because the element is static. The wedge can be considered as a free body, and the vertical forces acting on it can be individually quantified. The forces acting on the free body are gravitational force, the vertical pressure force on plane *a*, and the vertical component of the pressure force on plane *c*.

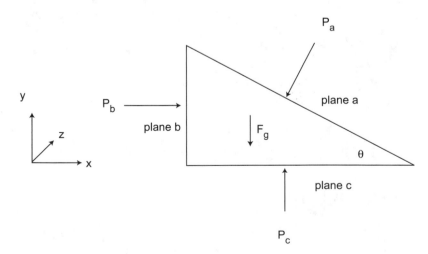

Figure 3-4. Forces on a wedge-shaped fluid volume.

Gravitational Force, F_g

From Newton's second law, the gravitational force of the wedge, F_g, is its mass (volume multiplied by its fluid density) multiplied by the constant of gravitational acceleration, g:

$$F_g = m \cdot \frac{g}{g_c} = \left(\rho \cdot \frac{\Delta x \cdot \Delta y}{2} \cdot \Delta z \right) \frac{g}{g_c} \tag{3-2}$$

Note that Newton's law proportionality factor, g_c, is needed when using Newton's second law with the U.S. Customary unit system, which defines mass, length, time, and force independently. Its value is $32.174 \text{ ft} \cdot \text{lbm/lbf} \cdot \text{s}^2$. When using the SI system of units, g_c is not needed because only mass, length, and time are independently defined and force units are defined in relation to mass, length, and time units ($1 \text{ N} \equiv 1 \text{ kg} \cdot \text{m/s}^2$). For completeness, g_c is included in the equation for Newton's second law.

The Vertical Pressure Force on Plane a, $F_{P,y(a)}$

The force on plane a from fluid pressure, $F_{P,y(a)}$, is the pressure in the y direction multiplied by the area of plane a:

$$F_{P,y(a)} = P_a \cdot A_a = P_a \cdot \frac{\Delta x}{\cos \theta} \cdot \Delta z \tag{3-3}$$

where P_a is the local pressure on plane a and A_a is the area in plane a on which the pressure acts.

The Vertical Component of the Pressure Force on Plane c, $F_{P,y(c)}$

The vertical component of the pressure force on plane c is

$$F_{P,y(c)} = P_c \cdot A_c = P_c \cdot (\Delta x \cdot \Delta z) \tag{3-4}$$

Because the fluid element is static, the sum of the forces must be equal to zero:

$$\Sigma F = 0 \tag{3-5}$$

Accounting for the forces acting on the free body gives the force balance of

$$-F_g + F_{P,y(c)} - F_{P,y(a)} = 0 \tag{3-6}$$

Note the minus signs on F_g and $F_{P,y(a)}$ account for the direction of the forces, as the positive y direction is upward.

Putting in the terms for the individual forces that were just derived, we obtain

$$-\left(\rho \cdot \frac{\Delta x \cdot \Delta y}{2} \cdot \Delta z\right)\frac{g}{g_c} + P_c \cdot (\Delta x \cdot \Delta z) - P_a \cdot \frac{\Delta x}{\cos \theta} \cdot \Delta z = 0 \qquad (3\text{-}7)$$

Dividing by Δx and Δz and rearranging yields

$$-\left(\rho \cdot \frac{\Delta y}{2}\right)\frac{g}{g_c} + P_c - P_a \cdot \frac{1}{\cos \theta} = 0 \qquad (3\text{-}8)$$

Take the limit $\Delta\theta \rightarrow 0$. This physically forces Δy to approach 0, and $\cos \theta$ approaches 1. The first term in the equation is then 0, and the equation becomes

$$P_c - P_a = 0 \qquad (3\text{-}9)$$

which can be rearranged to

$$\Rightarrow P_a = P_c \qquad (3\text{-}10)$$

Similarly, a force balances can also be written in the x direction, and it can be found that

$$\therefore P = P_a = P_b = P_c \qquad (3\text{-}11)$$

This proof shows that pressure at a point in a fluid is independent of direction. That is, static pressure acts equally in all directions at a point in a fluid.

Defining Pressure Datums

Pressure can be defined relative to two datums: atmospheric pressure or absolute zero pressure. As illustrated in Fig. 3-5, when pressure is defined relative to atmospheric pressure, it is termed "gauge pressure" and the units are usually pounds per square inch gauge (psig) or Pa gauge. So if a zero gauge pressure is indicated, then the pressure is atmospheric. If pressure is indicated to be "10 psig," then the pressure is 10 psi above atmospheric pressure. Pressure may also be defined relative to absolute zero pressure, in which case it is called absolute pressure. Absolute pressures are usually indicated with units of psia or Pa absolute. Pressure may also be reported as the difference between two values, called differential pressure (psid). Pressure gauges are designed to indicate gauge, absolute, or differential pressure, and it is important to know which datum the gauge is using. Vacuum gauges read negative pressures relative to atmospheric pressure (in positive values). The datum used for pressure should always be specified. Additional details of pressure gauges are given in Chapter 9.

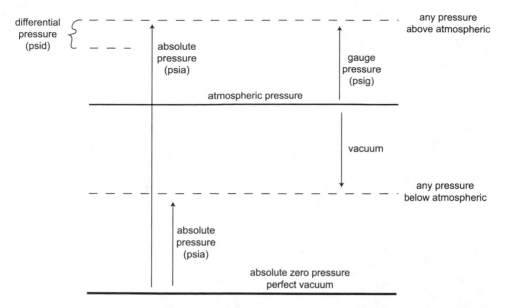

Figure 3-5. Pressure datums.

Variation of Pressure with Elevation in a Fluid Column

The pressure within a static fluid column changes with depth of the fluid. It is important for an engineer to be able to quantify pressure changes resulting from static height differences. The following is a derivation of an equation for calculating pressure change with change in fluid depth.

Consider a cylinder of fluid of cross-sectional area A and height Δy from within the bulk fluid as shown in Fig. 3-6. The force resulting from fluid pressure on the top surface, at $y = y + \Delta y$, is

$$F_P \big|_{y+\Delta y} = P \big|_{y+\Delta y} \cdot A \tag{3-12}$$

The force resulting from fluid pressure on the bottom surface, at $y = y$, is

$$F_P \big|_{y} = P \big|_{y} \cdot A \tag{3-13}$$

The force resulting from gravity, F_g, is

$$F_g = A \cdot \Delta y \cdot \rho \cdot \frac{g}{g_c} \tag{3-14}$$

Because the fluid cylinder is static, the sum of the forces on the cylinder must be equal to zero. Summing the forces acting on the cylinder in the vertical (y) direction yields

$$F_P \big|_{y} - F_P \big|_{y+\Delta y} - F_g = 0 \tag{3-15}$$

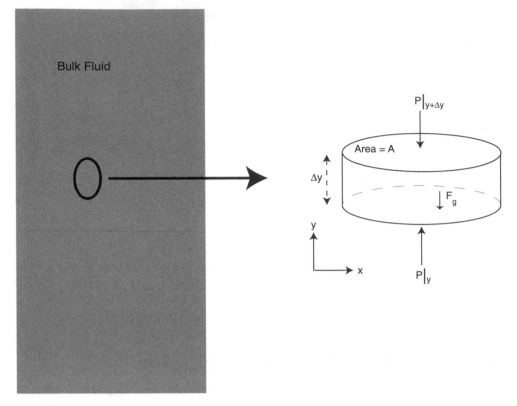

Figure 3-6. Forces on a fluid cylinder.

Inserting the individual force terms just derived gives the force balance:

$$(P|_y \cdot A) - (P|_{y + \Delta y} \cdot A) - \left(A \cdot \Delta y \cdot \rho \cdot \frac{g}{g_c} \right) = 0 \qquad (3\text{-}16)$$

Simplifying and rearranging yields

$$\frac{P|_{y + dy} - P|_y}{\Delta y} = -\rho \cdot \frac{g}{g_c} \qquad (3\text{-}17)$$

Letting $\Delta y \to 0$ yields the differential equation

$$\frac{dP}{dy} = -\rho \frac{g}{g_c} \qquad (3\text{-}18)$$

This is a fundamental equation in fluid statics that provides a relationship between static fluid pressure and fluid depth. The equation shows that static fluid pressure is related to the fluid density and the acceleration due to gravity, g.

The quantity $\rho g/g_c$ is called the specific weight γ of the fluid. For water at 20 °C (68 °F), the specific weight is calculated in SI units as

$$\gamma = \rho \frac{g}{g_c} = \left(998.21 \frac{\text{kg}}{\text{m}^3}\right) \frac{\left(9.81 \frac{\text{m}}{\text{s}^2}\right)}{(1)} = 9{,}790 \frac{\text{kg}}{\text{m}^2 \cdot \text{s}^2} = 9{,}790 \frac{\text{N}}{\text{m}^3}$$

In U.S. Customary units the specific weight is

$$\gamma = \rho \frac{g}{g_c} = \left(62.3 \frac{\text{lbm}}{\text{ft}^3}\right) \frac{\left(32.2 \frac{\text{ft}}{\text{s}^2}\right)}{\left(32.2 \frac{\text{ft} \cdot \text{lbm}}{\text{lbf} \cdot \text{s}^2}\right)} = 62.3 \frac{\text{lbf}}{\text{ft}^3}$$

Separating variables in the dP/dy expression (Eq. 3-18) gives

$$dP = -\rho \frac{g}{g_c} dy \tag{3-19}$$

Integrating from one height to another, we have

$$\int_{P_1}^{P_2} dP = -\int_{y_1}^{y_2} \rho \frac{g}{g_c} \cdot dy \tag{3-20}$$

Letting the fluid density remain constant (a common assumption made for liquids because they are considered incompressible) allows it to be moved outside the integral, giving

$$\int_{P_1}^{P_2} dP = -\rho \int_{y_1}^{y_2} \frac{g}{g_c} \tag{3-21}$$

The integration results in

$$P_2 - P_1 = -\rho \frac{g}{g_c}(y_2 - y_1) \tag{3-22}$$

This is the discretized form of Eq. 3-19 derived above and is useful for determining the change of pressure in a fluid from one depth to another when the fluid is not moving.

Example

If the water tank shown in the lower photograph in Fig. 3-2 contains 10-m-deep water at 12 °C (54 °F), the expected pressure at the bottom of the tank would be

$$P_2 - P_1 = -\rho \frac{g}{g_c}(y_2 - y_1) = -\left(999.5 \frac{\text{kg}}{\text{m}^3}\right)\frac{\left(9.81 \frac{\text{m}}{\text{s}^2}\right)}{(1)}\cdot(-10\ \text{m} - 0)$$

$$= 9.8 \times 10^4 \frac{\text{kg}}{\text{m} \cdot \text{s}^2} = 9.8 \times 10^4\ \text{Pa}$$

Note that y_1 is the reference height and set to 0 in the equation. The value for y_2 is negative as it is lower than the reference height. Since P_1 is atmospheric pressure, the pressure at the bottom of the tank is 9.8×10^4 Pa greater than atmospheric pressure. The pressure differential between the inside and outside of the tank is 9.8×10^4 Pa at the base of the tank. The greater the liquid column depth, the greater the pressure at the bottom of the tank will be. So a tall tower will provide high water pressure to a distribution system, yet the material in the structure will need to withstand the high stress from the pressure.

Example

A manometer contains two fluids at 68 °F (20 °C), water and an unknown fluid that is not miscible with water. The tops of both legs of the manometer are open to the atmosphere. As shown in Fig. 3-7, the "leg" containing the water has a static height of 8 in. above point B. The leg containing the unknown liquid has a static height of 9.25 in. above point C. Points B and C are at the same static height. What is the density of the unknown fluid?

Solution

The equation that quantifies the pressure difference between two points in a static fluid is

$$P_2 - P_1 = -\rho \frac{g}{g_c}(y_2 - y_1)$$

Writing the equation between points A and B gives

$$P_B - P_A = -\rho_{\text{fluid 1}} \frac{g}{g_c}(y_B - y_A)$$

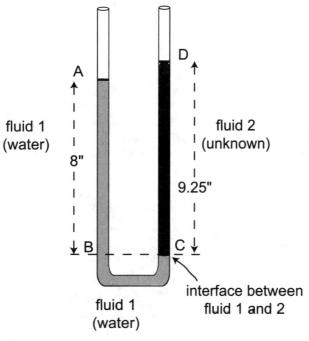

Figure 3-7. Two fluids in a manometer.

Rearranging to solve for P_B then gives

$$P_B = P_A - \rho_{\text{fluid 1}} \frac{g}{g_c} (y_B - y_A)$$

The pressure at points B and C is the same:

$$P_B = P_C$$

Combining with the previous expression, we have

$$P_C = P_A - \rho_{\text{fluid 1}} \frac{g}{g_c} (y_B - y_A)$$

and the pressure between points C and D is

$$P_D - P_C = -\rho_{\text{fluid 2}} \frac{g}{g_c} (y_D - y_C)$$

Rearranging yields

$$P_C = P_D + \rho_{\text{fluid 2}} \frac{g}{g_c}(y_D - y_C)$$

Combining the two equations for P_C yields

$$P_A - \rho_{\text{fluid 1}} \frac{g}{g_c}(y_B - y_A) = P_D + \rho_{\text{fluid 2}} \frac{g}{g_c}(y_D - y_C)$$

Pressures P_A and P_D are equal (atmospheric pressure) and can be subtracted from both sides of the expression, and the g/g_c term cancels from both sides by dividing by g/g_c:

$$-\rho_{\text{fluid 1}}(y_B - y_A) = \rho_{\text{fluid 2}}(y_D - y_C)$$

Rearranging gives

$$\rho_{\text{fluid 2}} = -\rho_{\text{fluid 1}} \frac{(y_B - y_A)}{(y_D - y_C)}$$

Plugging in values to find the density of fluid 2, we obtain

$$\rho_{\text{fluid 2}} = -62.3 \frac{\text{lbm}}{\text{ft}^3} \cdot \frac{(-8 \text{ in.})}{(9.25 \text{ in.})} = 53.9 \frac{\text{lbm}}{\text{ft}^3}$$

Static Pressure Forces on Surfaces

Fluid pressure exerts forces on solid surfaces with which they are in contact. The magnitude of the forces from the fluid pressure must be understood so that the solid surfaces do not rupture. Although pressure is independent of direction (as shown earlier), it exerts a force on a solid normal to the surface as shown in Fig. 3-8.

For a differential surface area in contact with a fluid under pressure, such as in Fig. 3-8, the force exerted by the fluid on the differential area dA at a given pressure P is

$$dF = P \cdot dA \qquad\qquad (3\text{-}23)$$

For a surface where the pressure does not vary over the surface area, this equation can be integrated to

$$\int dF = P \cdot \int dA \rightarrow F = P \cdot A \qquad\qquad (3\text{-}24)$$

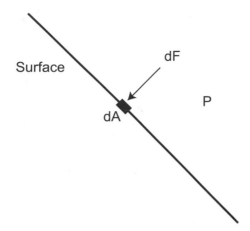

Figure 3-8. Differential force on a surface.

This equation can be used to calculate the force on a surface when the pressure does not change greatly with height, such as for gases with low density. It is also applicable for calculating the force on horizontal surfaces resulting from static liquid pressure. For a vertical surface in contact with a liquid with pressure differences over the contact area, the equation must be integrated.

Example

Rectangular gates are used in a channel (Fig. 3-9) to control and hold back wastewater. Each gate is 1.5 m wide and 4 m high. The gates are closed to completely stop the flow, which is at 10 °C. What is the force of the static water on each gate

Figure 3-9. Pressure force on a gate.

if the water rises to the top of the gates on the upstream side, with no water on the downstream side?

Solution

To determine the pressure on the gate, we start with

$$dF = P \cdot dA$$

Integrating, we get

$$\int dF = \int P \cdot dA$$

and so

$$F = \int P \cdot dA$$

Substituting in the barometric equation for P and $W \cdot dh$ for dA (both functions of depth) gives

$$F = \int \left(P_{atm} + \rho h \frac{g}{g_c} \right) \cdot W \cdot dh$$

where W is the width of the gate and dh is differential height.

This equation can be rearranged to

$$F = \int P_{atm} W \cdot dh + \int p \frac{g}{g_c} W \cdot h \cdot dh$$

On the upstream, submerged side of the gate, we can integrate from the lower edge of the gate (height = 0) to the top edge of the gate (height = h), so

$$F_{upstream} = P_{atm} W \int_0^h dh + \rho \frac{g}{g_c} W \int_0^h h \cdot dh$$

$$= P_{atm} W \left. h \right|_0^h + \rho \frac{g}{g_c} W \left. \frac{h^2}{2} \right|_0^h$$

$$= P_{atm} W h + \rho \frac{g}{g_c} W \frac{h^2}{2}$$

Similarly, the force on the downstream side of the gate is

$$F_{downstream} = P_{atm} W h + \rho \frac{g}{g_c} W \frac{h^2}{2}$$

However, since the downstream side is empty, the pressure on the downstream side of the gate results only from atmospheric pressure (ρ of air is small relative to that of water), so the second term is negligible, and so

$$F_{\text{downstream}} = P_{\text{atm}} Wh$$

Therefore the net force on the gate is

$$F_{\text{net}} = F_{\text{upstream}} - F_{\text{downstream}}$$

$$F_{\text{net}} = \left[P_{\text{atm}} Wh + \rho \frac{g}{g_c} W \frac{h^2}{2} \right] - \left[P_{\text{atm}} Wh \right]$$

$$= \rho \frac{g}{g_c} W \frac{h^2}{2}$$

$$= \left(999.7 \frac{\text{kg}}{\text{m}^3} \right) \frac{9.81 \frac{\text{m}}{\text{s}^2}}{1 \frac{\text{kg} \cdot \text{m}}{\text{N} \cdot \text{s}^2}} (1.5\text{m}) \frac{(4 \text{ m})^2}{2}$$

$$= 1.2 \times 10^5 \, \text{N}$$

Pressure Forces on Curved Surfaces

The pressure force on a curved surface that is submerged in a liquid is a little more complicated. Figure 3-10 illustrates pressure on a curved surface (adapted from Wilkes 1999). The differential area in the right side of Fig. 3-10 shows the local pressure force acting normal to the surface. Starting with

$$F = \int P \cdot dA$$

we have for the horizontal component of the pressure force

$$F = \int (P \cdot \sin \theta) \cdot dA \tag{3-25}$$

where θ is defined in the figure. The area of the curve projected onto the vertical axis is

$$dA' = dA \cdot \sin \theta \tag{3-26}$$

Upon substitution, we see that the horizontal pressure force is a function of the pressure and the projected area:

$$F = \int P \cdot dA' \tag{3-27}$$

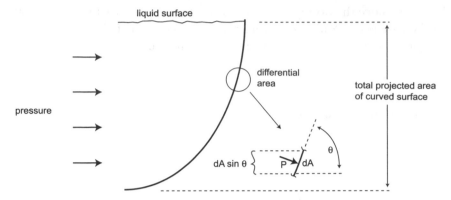

Figure 3-10. Pressure force on a curved surface. *Source*: Adapted from Wilkes (1999).

Example

(Adapted from Wilkes 1999) A 50-m-long, 1-m-diameter pipe has one side (external) completely covered with water at 15 °C, with no water on the other side, as shown in Fig. 3-11. What is the horizontal force on the pipe?

Solution

Starting with the equation just derived, we have

$$F_x = \int P \cdot dA'$$

The pressure in the water is a function of depth:

$$P_2 - P_1 = -\rho \frac{g}{g_c}(y_2 - y_1)$$

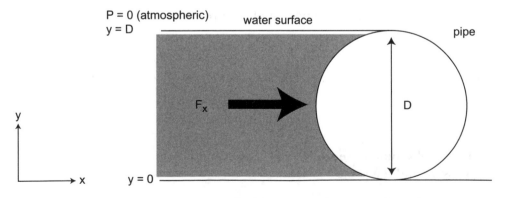

Figure 3-11. Horizontal pressure force on a pipe with water on one side.

Let $y_1 = y$, $y_2 = D$, $P_1 = P$, and $P_2 = 0$. Note the bottom of the pipe is $y = 0$. Then

$$0 - P = -\rho \frac{g}{g_c}(D - y)$$

Rearranging gives

$$P = \rho \frac{g}{g_c}(D - y)$$

and the projected area is

$$dA' = L \cdot dy$$

where L is the pipe length. Combining these expressions we get

$$F_x = \int_0^D \rho \frac{g}{g_c} L(D - y) \cdot dy$$

Simplifying, we obtain

$$F_x = \rho \frac{g}{g_c} L \int_0^D (D - y) \cdot dy$$

which integrates to

$$F_x = \rho \frac{g}{g_c} L \left(\frac{1}{2} D^2 \right)$$

Entering in the numerical values, we have

$$F_x = \left(999.1 \frac{kg}{m^3} \right) \frac{9.81 \frac{m}{s^2}}{1 \frac{kg \cdot m}{N \cdot s^2}} (50 \text{ m}) \frac{(1 \text{ m})^2}{2}$$

and so

$$F_x = 2.4 \times 10^5 \text{ N}$$

Symbol List

A	area
D	depth
F	force
F_n	normal force
g	constant of gravitational acceleration
g_c	Newton's law proportionality factor, 32.174 ft·lbm/lbf·s^2 for the U.S. Customary System of units
h	height
L	length
m	mass
P	pressure
W	width
γ	specific weight
θ	angle
ρ	density

Problems

1. A water distribution pipe runs from a drinking water treatment plant in a valley to a point for distribution 2,000 ft higher in elevation than the valley. If the water in the line does not flow, what is the water pressure in psi in the line at the distribution point (at 2,000 ft elevation) in relation to the pressure at the drinking water plant? The water is at a temperature of 10 °C.

2. A dam contains water that is 150 ft above the top of a 10-ft square gate. The gate can be opened or closed to let water escape from the impoundment created by the dam. What is the total force on the gate?

3. Gasoline has contaminated a groundwater aquifer. Pumps are used to remove the gasoline and water mixture and the mixture is being allowed to separate into the two phases (gasoline and water) in a tank that is 3 m in depth, as shown in Fig. 3-12. The tank has two sight glasses to indicate the level of gasoline and the level of water. After the tank is filled with the mixture, and a sufficient amount of time for phase separation has occurred, the water level in the sight glass is 2.5 m above the bottom of the tank. What is the gasoline/water interface height in the tank (above the tank bottom)? Assume the temperature is 10 °C and the density of the gasoline phase is 737 kg/m^3.

4. Water is stored in a closed (to the atmosphere) tank. The water is 12 ft above the bottom of the tank and the air in the tank is pressurized to 55 psig, as shown in Fig. 3-13. What is the pressure at the bottom of the tank? Assume the temperature of the water is 50 °F.

5. The tank in Problem 4 has a round access opening in its bottom of for maintenance purposes. The opening is 24 in. in diameter. The cover is fastened with 10 fasteners. Assuming the fasteners are equally loaded, what is the force on each fastener?

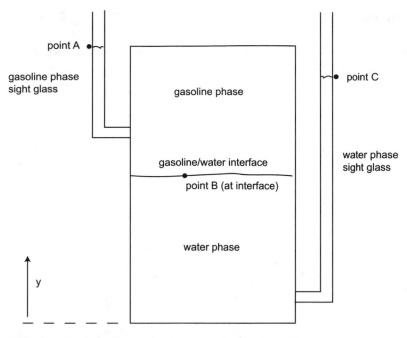

Figure 3-12. Tank for gasoline/water separation in Problem 3.

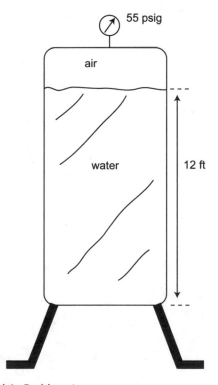

Figure 3-13. Tank in Problem 4.

References

Aqueduct Commission, NY (1907). Reports on the New Croton Aqueduct, Reservoirs and Dams, 1895 to 1907.

Wilkes, J. O. (1999). *Fluid Mechanics for Chemical Engineers*, Prentice Hall, Upper Saddle River, NJ.

CHAPTER 4

Fundamentals of Fluid Flow

Chapter Objectives

1. Assemble mass balances for treatment systems.
2. Recall derivation of the equations of motion.
3. Identify various energy forms and formulate energy balances.
4. Solve Bernoulli's equation for treatment systems.

General Balances

An important concept in all branches of engineering is that of a balance. A balance is simply a proper accounting of an item of interest, such as energy, momentum, total mass, or mass of an individual species (such as a contaminant). More precisely, a balance is a mathematical statement describing the change in quantity of an item of interest within defined system boundaries by transport, reaction, accumulation, etc.

A balance analogy with which we are all familiar is balancing our checkbook. If we are not careful to account for the money deposited and withdrawn from our checking account, we can have an overdrawn account or "bounce" a check. Some of us know the problems with bouncing a check. Conversely, having a high accumulation of money in a checking account may not be a good idea, as a better interest rate (for those checking accounts that are paying an interest rate) can be found in other types of accounts. So we all are probably aware of the need to carefully account for the money coming into our accounts, the money leaving, and the form in which the money exists (savings versus checking accounts). In engineering, many times we will want to account for energy, mass, etc. The simple accounting of mass and energy in a system is often the only way to understand the operation or predict the performance of a treatment system. Moreover, an incorrect accounting of mass and energy in a treatment system can have very dire consequences. This chapter discusses the general approach for conducting a balance, and then focuses on mass, momentum, and energy balances for treatment systems.

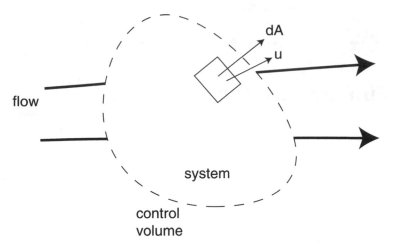

Figure 4-1. A system showing transport (flow) through a control volume.

A balance is written for a specific entity of interest over a system boundary, or control volume, as shown in Fig. 4-1.

Balances can be conducted on *extensive* properties—properties that are dependent on the amount of material present. Energy, mass, and momentum are properties that are dependent on the amount of material present, and they are therefore extensive properties. *Intensive* properties are independent of the amount of material. Density, viscosity, and pressure are intensive properties, and balances cannot be made on them.

Many balances can be performed for treatment systems; these include the following:

1. overall material or mass balances,
2. individual component material balances over various stages of the process,
3. momentum balances, and
4. energy balances.

Mass Balances

Mass balances are constructed based on the principle of conservation of mass. That is, mass cannot be created or destroyed, only changed in form. The amount of mass in a closed system remains constant.

A mass balance is a mathematical statement describing the change in mass of a constituent within defined system boundaries (the control volume) by transport across boundaries and/or by reaction within system boundaries. It may be written as

$$\begin{pmatrix} \text{input through} \\ \text{system boundaries} \end{pmatrix} + \begin{pmatrix} \text{generation} \\ \text{inside system} \end{pmatrix} = \begin{pmatrix} \text{output through} \\ \text{system boundaries} \end{pmatrix} + \text{accumulation} \quad (4\text{-}1)$$

The first and third terms account for the mass that is transported into and out of the control volume, respectively. The second term accounts for generation within the control volume. In hydraulics calculations for treatment systems, we can usually assume that there is no material generation, so the second term is usually unnecessary. It is left here to keep the balance in general terms. The fourth term accounts for accumulation of mass within the control volume, which may increase or decrease for transient processes. For processes that are steady state (defined as occurring when the state of mass throughout the system, at every point in the control volume, does not change with time) and steady flow (in which there is constant mass flux through the system boundaries), accumulation in the control volume is zero.

There are some general rules for constructing mass balances:

1. Draw a flow diagram of the process and choose a control volume.
2. Place all available data on the diagram.
3. Determine flow rates and material compositions (concentrations or densities).
4. Select a basis for calculations (such as pounds or grams). Make sure the units used are consistent.

Mass balances can be accomplished on two spatial scales:

1. microscale, with a differential "point" form, or
2. macroscale, with a discretized form.

Microscale Mass Balance

If we consider a small element in space, such as shown in the Fig. 4-2, and write a mass balance in three dimensions, we can get the "point" form of the mass balance equation. The mass in a cube of dimensions Δx, Δy, and Δz and density ρ at time t is

$$\rho|_t \, \Delta x \Delta y \Delta z \tag{4-2}$$

and at time $t + \Delta t$ the mass is

$$\rho|_{t+\Delta t} \, \Delta x \Delta y \Delta z \tag{4-3}$$

So the accumulation over Δt is

$$(\rho|_{t+\Delta t} \, \Delta x \Delta y \Delta z) - (\rho|_t \, \Delta x \Delta y \Delta z) \tag{4-4}$$

The flow into the cube through face a is

$$\rho_a \cdot V_{x_a} \cdot \Delta y \cdot \Delta z \cdot \Delta t \tag{4-5}$$

where V_{x_a} is the velocity in the x direction through face a due to the flow.

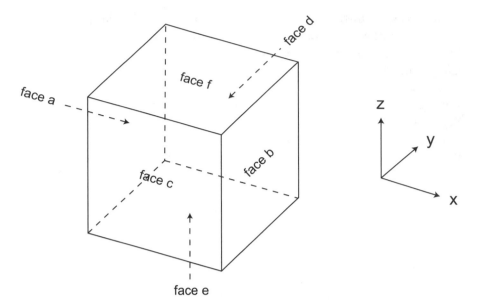

Figure 4-2. Fluid element for three-dimensional mass balance.

The flow out of the cube through face b is

$$\rho_b \cdot V_{x_b} \cdot \Delta y \cdot \Delta z \cdot \Delta t \qquad (4\text{-}6)$$

where V_{x_b} is the velocity in the x direction through face b.

The flow through the other faces can be written similarly and substituted into the mass balance equation

$$\begin{pmatrix} \text{input through} \\ \text{system boundaries} \end{pmatrix} + \begin{pmatrix} \text{generation inside system} \end{pmatrix} = \begin{pmatrix} \text{output through} \\ \text{system boundaries} \end{pmatrix} + \text{accumulation}$$

to yield

$$[(\rho_a V_{x_a} \Delta y \Delta z \Delta t) + (\rho_c V_{y_c} \Delta x \Delta z \Delta t) + (\rho_e V_{z_e} \Delta x \Delta y \Delta t)] + 0$$

$$= [(\rho_b V_{x_b} \Delta y \Delta z \Delta t) + (\rho_d V_{y_d} \Delta x \Delta z \Delta t) + (\rho_f V_{z_f} \Delta x \Delta y \Delta t)] \qquad (4\text{-}7)$$

$$+ [(\rho|_{t+\Delta t} - \rho|_t) \Delta x \Delta y \Delta z]$$

Rearranging yields

$$-\frac{(\rho|_{t+\Delta t} - \rho|_t)}{\Delta t} \Delta x \Delta y \Delta z = [(\rho_b V_{x_b} \Delta y \Delta z) + (\rho_d V_{y_d} \Delta x \Delta z) + (\rho_f V_{z_f} \Delta x \Delta y)]$$

$$- [(\rho_a V_{x_a} \Delta y \Delta z) + (\rho_c V_{y_c} \Delta x \Delta z) + (\rho_e V_{z_e} \Delta x \Delta y)] \qquad (4\text{-}8)$$

Simplifying, we have

$$-\frac{(\rho|_{t+\Delta t} - \rho|_t)}{\Delta t} = \left[\left(\frac{\rho_b V_{x_b} - \rho_a V_{x_a}}{\Delta x} \right) + \left(\frac{\rho_d V_{y_d} - \rho_c V_{y_c}}{\Delta y} \right) + \left(\frac{\rho_f V_{z_f} - \rho_e V_{z_e}}{\Delta z} \right) \right] \quad (4\text{-}9)$$

Taking the limits $\Delta t \rightarrow 0$, $\Delta x \rightarrow 0$, $\Delta y \rightarrow 0$, and $\Delta z \rightarrow 0$ then yields

$$-\frac{\partial \rho}{\partial t} = \frac{\partial(\rho V_x)}{\partial x} + \frac{\partial(\rho V_y)}{\partial y} + \frac{\partial(\rho V_z)}{\partial z} \quad (4\text{-}10)$$

This is the three-dimensional *continuity equation* at a point in the Cartesian coordinate system.

Macroscale Mass Balance

Consider a nonreacting incompressible flow through the steady-state and steady-flow system shown in Fig. 4-3 with a varying cross section. Mass flows into the system at boundary 1. So the net input to the system through the system boundary is

$$\rho_1 \cdot A_1 \cdot V_1 \cdot \Delta t \quad (4\text{-}11)$$

where A_1 is the area at boundary 1 and V_1 is the fluid velocity at boundary 1. The output of mass from the system through boundary 2 is

$$\rho_2 \cdot A_2 \cdot V_2 \cdot \Delta t \quad (4\text{-}12)$$

where A_2 is the area at boundary 2 and V_2 is the fluid velocity at boundary 2.

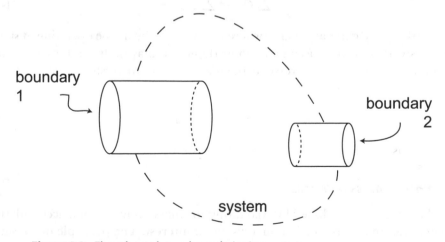

Figure 4-3. Flow through two boundaries in a system.

In this system there is no generation or accumulation of the fluid. So the mass balance becomes

$$\rho_1 A_1 V_1 \Delta t + 0 = \rho_2 A_2 V_2 \Delta t + 0 \tag{4-13}$$

The time increment, Δt, cancels, yielding

$$\rho_1 A_1 V_1 = \rho_2 A_2 V_2 \tag{4-14}$$

The quantity $\rho A V$ is the mass flow rate, $\dot{m}$:

$$\dot{m} = \rho A V \tag{4-15}$$

Units for mass flow rate include kilograms per second and pounds-mass per second.

For constant density, Eq. 4-14 simplifies to

$$A_1 V_1 = A_2 V_2 \tag{4-16}$$

The quantity AV is the volumetric flow rate, Q:

$$Q = AV \tag{4-17}$$

Units for Q include cubic meters per second, cubic feet per second, and gallons per minute (gpm).

Then, for this system, which is assumed to be incompressible, with no mass accumulation or generation, we have

$$Q_1 = Q_2 \tag{4-18}$$

For a system with multiple inflows and outflows and an incompressible fluid, a more general equation is

$$\sum Q_{in} = \sum Q_{out} \tag{4-19}$$

For the more general scenario where the system may not be operating at steady state or steady flow (i.e., accumulation or depletion may occur in the system) and the fluid may not be incompressible, the mass input to the system is $(\dot{m}_{in} \cdot \Delta t)$, and the output is $(\dot{m}_{out} \cdot \Delta t)$:

$$(\dot{m}_{in} \cdot \Delta t) + 0 = (\dot{m}_{out} \cdot \Delta t) + \Delta m \tag{4-20}$$

where Δm is the change in mass within the system.

Summary of Mass Balances

Mass balances are developed by accounting for mass input, output, accumulation, and generation (where needed) in a system of interest. The principle of conservation of mass is an important concept applied in all branches of engineering, including hydraulics and treatment systems. The resulting mass balance applied to a system is also called the continuity equation, and one often refers to this as just continuity.

Example

Equalization tanks are frequently used in treatment systems to provide a more uniform flow rate to following treatment processes, to "dampen" flow variations to a certain extent. In this example, wastewater is flowing into an equalization tank at the head of a treatment system. See Fig. 4-4. The wastewater flows at 16,000 gpm for 6 hr, and the system treats 14,000 gpm maximum. If the equalization tank starts with 100,000 gallons in it, what volume of the wastewater is in the tank at the end of 6 hr?

Solution

From Eq. 4-20, we have

$$(\dot{m}_{in} \cdot \Delta t) = (\dot{m}_{out} \cdot \Delta t) + \Delta m$$

The fluid density can be assumed to be constant, so we divide by ρ to get

$$\left(\frac{\dot{m}_{in}}{\rho} \cdot \Delta t \right) = \left(\frac{\dot{m}_{out}}{\rho} \cdot \Delta t \right) + \frac{\Delta m}{\rho}$$

which simplifies to

$$\Delta v = (Q_{in} - Q_{out})(\Delta t)$$

Substituting numerical values then gives

$$\Delta v = \left(16,000 \frac{\text{gal}}{\text{min}} - 14,000 \frac{\text{gal}}{\text{min}} \right)(6 \text{ hr})\left(\frac{60 \text{ min}}{1 \text{ hr}} \right)$$

$$= 720,000 \text{ gal}$$

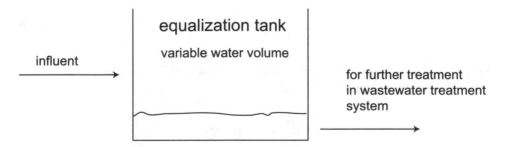

Figure 4-4. Equalization tank example.

So the total volume of wastewater in the equilibration tank at the end of 6 hr is 720,000 gal (the net inflow) plus the initial 100,000 gal, or 820,000 gal total.

Equations of Motion

Newton's second law of motion states

$$\text{mass} \cdot \text{acceleration} = \sum \text{forces} \qquad (4\text{-}21)$$

For fluids, the mass of a fluid element in motion possesses momentum. Momentum per unit volume (similar to mass per unit volume or ρ), or momentum density, is defined as

$$\text{momentum density} \equiv \rho \cdot \text{velocity} \qquad (4\text{-}22)$$

Dividing each term in Newton's second law by volume **v** yields

$$\frac{\text{mass}}{\textbf{v}} \cdot \text{acceleration} = \sum \frac{\text{forces}}{\textbf{v}} \qquad (4\text{-}23)$$

or

$$\rho \cdot \text{acceleration} = \sum \vec{f} \qquad (4\text{-}24)$$

where $\vec{f}$ represents the forces per unit volume (a vector) exerted on the fluid element body or fluid element surfaces, and $\sum \vec{f}$ is the net force per unit volume.

For a fluid element moving along a streamline of fluid flow, the velocity of the fluid element is a function of time and location. The acceleration of the fluid element can be found by using the chain rule of differentiation:

$$\text{acceleration} = \frac{\partial \vec{V}}{\partial t} + \left(\frac{\partial \vec{V}}{\partial x}\frac{dx}{dt} + \frac{\partial \vec{V}}{\partial y}\frac{dy}{dt} + \frac{\partial \vec{V}}{\partial z}\frac{dz}{dt} \right) \qquad (4\text{-}25)$$

Since $dx/dt = V_x$, $dy/dt = V_y$, and $dz/dt = V_x$, we can write

$$\text{acceleration} = \frac{\partial \vec{V}}{\partial t} + \left(V_x \frac{\partial \vec{V}}{\partial x} + V_y \frac{\partial \vec{V}}{\partial y} + V_z \frac{\partial \vec{V}}{\partial z} \right) \qquad (4\text{-}26)$$

Note that acceleration can also be written as the *material* or *substantial derivative* of velocity, where the material derivative is

$$\frac{D}{Dt} = \frac{\partial}{\partial t} + (\vec{V} \cdot \nabla)$$

The vector differential operator ∇ ("del" or "nabla") in rectangular coordinates is

$$\nabla = \sum_i \delta_i \frac{\partial}{\partial x_i}$$

where x_i are variables associated with the axes and δ_i are the unit vectors. So Newton's second law becomes

$$\rho \cdot \frac{D\vec{V}}{Dt} = \sum \vec{f}$$

Inserting the term for acceleration defined by Eq. 4-26 into Eq. 4-24 for acceleration gives the *general momentum balance*:

$$\rho \left[\frac{\partial \vec{V}}{\partial t} + \left(V_x \frac{\partial(\vec{V})}{\partial x} + V_y \frac{\partial(\vec{V})}{\partial y} + V_z \frac{\partial(\vec{V})}{\partial z} \right) \right] = \sum \vec{f} \qquad (4\text{-}27)$$

Note that V_x, V_y, and V_z are the components of local fluid velocity, $\vec{V}$ (a vector).

So, starting with our general balance approach, we constructed a momentum balance. But the general balance must be modified by accounting for forces exerted on the fluid element. So the word definition of the *momentum balance* becomes

$$\begin{pmatrix} \text{net rate of flow of} \\ \text{momentum through} \\ \text{system boundaries} \end{pmatrix} + \begin{pmatrix} \text{sum of forces on} \\ \text{fluid boundaries} \end{pmatrix} = \begin{pmatrix} \text{accumulation} \\ \text{(rate of change of} \\ \text{momentum in system)} \end{pmatrix} \qquad (4\text{-}28)$$

Rearranging this we get

$$\begin{pmatrix} \text{accumulation} \\ \text{(rate of change of} \\ \text{momentum in system)} \end{pmatrix} = \begin{pmatrix} \text{net rate of flow of} \\ \text{momentum through} \\ \text{system boundaries} \end{pmatrix} + \begin{pmatrix} \text{sum of forces on} \\ \text{fluid boundaries} \end{pmatrix} \qquad (4\text{-}29)$$

This word definition of the conservation of momentum matches the equation defining the momentum balance, Eq. 4-27, where

$$\text{accumulation} = \rho \frac{\partial \vec{V}}{\partial t}$$

$$\text{net rate of momentum through system boundaries} = \rho \left(V_x \frac{\partial(\vec{V})}{\partial x} + V_y \frac{\partial(\vec{V})}{\partial y} + V_z \frac{\partial(\vec{V})}{\partial z} \right)$$

$$\text{sum of forces on fluid boundaries} = \sum \vec{f}$$

A fluid element may be affected by both body forces (distributed throughout the fluid element) and surface forces. The force of gravity on the fluid in the element is a body force. Normal (σ) and shear stresses (τ) are due to surface forces. Incorporating these forces, we can write the momentum equation in the x, y, and z directions to form the *equations of motion* in Cartesian coordinates (see Table 4-1).

Table 4-1. Equations of motions and the Navier–Stokes equations.

Equations of Motion (for Cartesian coordinate system)

x direction:

$$\rho\left[\frac{\partial V_x}{\partial t}+\left(V_x\frac{\partial(V_x)}{\partial x}+V_y\frac{\partial(V_x)}{\partial y}+V_z\frac{\partial(V_x)}{\partial z}\right)\right]=\rho g_x+\left[\frac{\partial\sigma_{xx}}{\partial x}+\frac{\partial\tau_{yx}}{\partial y}+\frac{\partial\tau_{zx}}{\partial z}\right]$$

y direction:

$$\rho\left[\frac{\partial V_y}{\partial t}+\left(V_x\frac{\partial(V_y)}{\partial x}+V_y\frac{\partial(V_y)}{\partial y}+V_z\frac{\partial(V_y)}{\partial z}\right)\right]=\rho g_y+\left[\frac{\partial\tau_{xy}}{\partial x}+\frac{\partial\sigma_{yy}}{\partial y}+\frac{\partial\tau_{zy}}{\partial z}\right]$$

z direction:

$$\rho\left[\frac{\partial V_z}{\partial t}+\left(V_x\frac{\partial(V_z)}{\partial x}+V_y\frac{\partial(V_z)}{\partial y}+V_z\frac{\partial(V_z)}{\partial z}\right)\right]=\rho g_z+\left[\frac{\partial\tau_{xz}}{\partial x}+\frac{\partial\tau_{yz}}{\partial y}+\frac{\partial\sigma_{zz}}{\partial z}\right]$$

Navier–Stokes Equations

x direction:

$$\rho\left[\frac{\partial V_x}{\partial t}+\left(V_x\frac{\partial(V_x)}{\partial x}+V_y\frac{\partial(V_x)}{\partial y}+V_z\frac{\partial(V_x)}{\partial z}\right)\right]=-\frac{\partial P}{\partial x}+\rho g_x+\mu\left[\frac{\partial^2 V_x}{\partial x^2}+\frac{\partial^2 V_x}{\partial y^2}+\frac{\partial^2 V_x}{\partial z^2}\right]$$

y direction:

$$\rho\left[\frac{\partial V_y}{\partial t}+\left(V_x\frac{\partial(V_y)}{\partial x}+V_y\frac{\partial(V_y)}{\partial y}+V_z\frac{\partial(V_y)}{\partial z}\right)\right]=-\frac{\partial P}{\partial y}+\rho g_y+\mu\left[\frac{\partial^2 V_y}{\partial x^2}+\frac{\partial^2 V_y}{\partial y^2}+\frac{\partial^2 V_y}{\partial z^2}\right]$$

z direction:

$$\rho\left[\frac{\partial V_z}{\partial t}+\left(V_x\frac{\partial(V_z)}{\partial x}+V_y\frac{\partial(V_z)}{\partial y}+V_z\frac{\partial(V_z)}{\partial z}\right)\right]=-\frac{\partial P}{\partial z}+\rho g_z+\mu\left[\frac{\partial^2 V_z}{\partial x^2}+\frac{\partial^2 V_z}{\partial y^2}+\frac{\partial^2 V_z}{\partial z^2}\right]$$

For incompressible Newtonian fluids, by substituting in terms for σ and τ (terms defined in pressure, viscosity, and rates of deformation) under laminar flow conditions, the *Navier–Stokes equations* can be derived from the momentum equation (see Table 4-1). The Navier–Stokes equations completely describe incompressible laminar Newtonian fluid flow. See Munson et al. (2002) for a derivation of the Navier–Stokes equations.

Thermodynamics

Although many contributed to the early knowledge of energy and its transformations (e.g., Carnot, Joule, Kelvin, and Helmholtz), the body of knowledge now known as thermodynamics is attributed in large part to the work of J. Willard Gibbs and Jacobus van't Hoff (Lewis and Randall 1923; Sposito 1981). Lewis and Randall (1923) stated that

> The fascination of a growing science lies in the work of the pioneers at the very borderland of the unknown, but to reach this frontier one must pass over well traveled roads; of these one of the safest and surest is the broad highway of thermodynamics.

This quote illustrates the importance and utility of thermodynamics in solving complex and unresolved problems. Thermodynamics is the science addressing the relationship among mechanical energy, work, and heat, and the conversions among them. It had its origins in steam engine design many years ago, but the subject now covers energy transfer and transformations of all kinds. A thermodynamic approach can be very useful for solving hydraulics problems in treatment systems. For our discussion of thermodynamics some definitions are in order:

System—control volume or element of space for study.
Surroundings—everything external to the system.
Boundary—the surface, either real or imaginary, that surrounds the system and may be either rigid or movable.
Boundary impermeable to mass flow—*closed system.*
Flow of at least one component—*open system.*
Extensive property—dependent on amount of material present (e.g., mass).
Intensive property—independent of the amount of material present (e.g., pressure or temperature).

There are four fundamental thermodynamic laws: the zeroth, first, second, and third laws of thermodynamics. Although all of the thermodynamic postulates are of fundamental importance, the first law of thermodynamics (also known as conservation of energy) is of particular utility for understanding hydraulic systems and will be discussed in detail later. It is the basis for Bernoulli's equation, one of the most useful equations for predicting the behavior of fluids.

As discussed earlier in the chapter, general balances can be made on extensive properties, such as mass. Balances can also be performed on energy, which is another extensive property. The fact that energy is conserved in all processes and systems and that an energy balance can be made is explained by the first law of thermodynamics: *Energy cannot be created nor destroyed, only converted in form.* The various energy forms with which we are concerned in treatment systems are discussed in the following section.

Typical units for energy in the SI system of units include joules [J], newton·meters [N·m], or kg·m^2·s^{-2}. In U.S. Customary units, one uses ft·lbf or lbm·ft·lbf^{-1}·s^{-2}. Other units used are the calorie and the British thermal unit [BTU].

Energy Forms

Internal Energy

A substance is composed of molecules that are in constant motion. Internal energy is the kinetic energy of translation and rotation of the molecules that makes up the substance. Therefore it is "internal" to the substance, and not a function of the substance's movement or location (as a whole). The translational and rotational motion of the molecules depends on temperature. A higher temperature produces greater motion of the molecules, and therefore the substance has an inherently greater internal energy.

Internal energy is a state function. That is, its value does not depend on the process taking place, just the present conditions (such as temperature and pressure). Absolute values of internal energy are not known; internal energy can only be quantified in comparison to a reference state. Absolute values of internal energy are not normally of concern; the changes in internal energy when going from one set of conditions to another are what matters.

The symbol U is used to denote internal energy and has energy units, whereas u is used for specific internal energy, or internal energy per unit mass.

Kinetic Energy

When a force is applied to a body, or an element of a substance, to accelerate it over a distance, work is done and can be quantified with

$$dW = F \cdot dx \qquad (4\text{-}30)$$

where F is the force causing a displacement dx of the body or element.

Applying Newton's second law, $F = ma$, gives

$$dW = ma \cdot dx \qquad (4\text{-}31)$$

By substituting the derivative of the velocity with time, dV/dt, for a, Eq. 4-29 becomes

$$dW = m\frac{dV}{dt} \cdot dx \qquad (4\text{-}32)$$

where V is the velocity of the body or element. Rearranging gives

$$dW = m\frac{dx}{dt} \cdot dV \qquad (4\text{-}33)$$

and since dx/dt = velocity, we have

$$dW = mV \cdot dV \qquad (4\text{-}34)$$

Integrating from V_1 to V_2 results in

$$W = m\int_{V_1}^{V_2} VdV = m\left(\frac{V_2^2}{2} - \frac{V_1^2}{2}\right) = \frac{1}{2}mV_2^2 - \frac{1}{2}mV_1^2 \qquad (4\text{-}35)$$

This equation shows that the work done on a body or element in accelerating it is equal to the change in the body's kinetic energy. The quantity $\frac{1}{2}mV^2$ is the *kinetic energy* that a body or element possesses owing to its motion at a given velocity V:

$$KE = \frac{1}{2}\frac{mV^2}{g_c} \qquad (4\text{-}36)$$

Note that as discussed in Chapter 3, Newton's law proportionality factor, g_c, must be included in the calculation for kinetic energy when using U.S. Customary units and not when using the SI system of units. It is included for completeness.

The specific kinetic energy, or kinetic energy per unit mass, is

$$ke = \frac{1}{2}\frac{V^2}{g_c} \qquad (4\text{-}37)$$

The specific kinetic energy may also be written as

$$ke = \frac{V^2}{2} \qquad (4\text{-}38)$$

which has units of m^2/s^2 or ft^2/s^2. Note that lowercase *ke* denotes *specific* kinetic energy versus uppercase for kinetic energy.

Potential Energy

Work is done when a force F is applied to a body or a "packet" or element of mass to raise it in elevation dz from a reference level:

$$dW = F \cdot dz \qquad (4\text{-}39)$$

The minimum force required to raise the body in elevation against gravity (opposite in direction from the direction of gravitational acceleration g) is

$$F = m \frac{g}{g_c} \qquad (4\text{-}40)$$

Substituting Eq. 4-40 into Eq. 4-39 for F yields

$$dW = m \frac{g}{g_c} \cdot dz \qquad (4\text{-}41)$$

Integrating from z_1 to z_2 gives

$$W = m \int_{z_1}^{z_2} \frac{g}{g_c} \cdot dz \qquad (4\text{-}42)$$

or

$$W = \Delta \frac{mgz}{g_c} \qquad (4\text{-}43)$$

The work done on a body or element in raising it in elevation is equal to the change in the body's potential energy. The quantity mgz/g_c is the potential energy that a body or element possesses because of its elevation:

$$PE = \frac{mgz}{g_c} \qquad (4\text{-}44)$$

Potential energy is the energy that a body has as a result of its location. Location can be, and usually is, its position in a gravitational field, but it can also be its location in another influencing field (i.e., a compressed spring or centrifugal force).

The specific potential energy, or potential energy per unit mass, is

$$pe = \frac{gz}{g_c} \qquad (4\text{-}45)$$

Sometimes the specific potential energy is written as

$$pe = gz \qquad (4\text{-}46)$$

with typical units of m^2/s^2 or ft^2/s^2. Note that lowercase pe denotes *specific* potential energy versus uppercase for potential energy.

Other Forms of Energy

Other forms of energy include electrostatic energy, magnetic energy, interfacial or surface energy, and nuclear energy. These other forms can be accounted for in energy balances by inclusion of additional terms if needed. For fluid systems, they are not normally needed but are mentioned in this discussion to be comprehensive. It is possible that new technologies will be developed in the future where these other forms of energies will need to be accounted for.

Energy Transfer

Energy can be transferred by the flow of *heat* from one element to another or from a system to its surroundings. Heat flows from one element to another, or from one "packet" of mass to another, through conduction or radiation, as well as convection, all of which are functions of temperature. It is the temperature difference that is the driving force for heat transfer from conduction and radiation.

Energy can also be transferred by *work*. This is work done on one body or element by another. Work (W) is a function of the magnitude of the force F exerted and the distance that the force operates:

$$W = \int F \cdot dx \tag{4-47}$$

This equation is for linear displacement. But a rotating shaft also does work (such as in a pump). The torque exerted by a rotating shaft is quantified by

$$T = F \cdot L \tag{4-48}$$

where F is the linear force applied at a distance L (the lever or moment arm) that produces torque T carried by a rotating shaft. Rearranging gives

$$F = \frac{T}{L} \tag{4-49}$$

The distance that the shaft moves as the torque is applied is

$$dx = L \cdot d\theta \tag{4-50}$$

where $d\theta$ is the differential angular displacement.

So the work done by a rotating shaft is quantified with

$$dW = F \cdot dx = \left(\frac{T}{L}\right)(L \cdot d\theta) \tag{4-51}$$

which simplifies to

$$dW = T \cdot d\theta \tag{4-52}$$

The power P transferred by or to a rotating shaft is then

$$P = \frac{dW}{dt} \tag{4-53}$$

Substituting Eq. 4-52 into Eq. 4-53 for dW gives

$$P = \frac{(T \cdot d\theta)}{dt} = T\frac{d\theta}{dt} = T\omega \tag{4-54}$$

where ω is the rotational velocity, which is 2π multiplied by the number of revolutions per unit time. Equation 4-53 can be rearranged to

$$dW = P \cdot dt \tag{4-55}$$

and Eq. 4-54 can be substituted into Eq. 4-55 for P to yield

$$dW = (T \cdot \omega) \cdot dt \tag{4-56}$$

Electrical work can be done also. A force on a charge in an electric field E can cause displacement of the charge. The force required to move a charge is

$$F = \varphi\frac{dE}{dx} \tag{4-57}$$

where φ is the charge and dE/dx is the potential (voltage) gradient. The work done in displacing a charge by dx is

$$W = \int F \cdot dx = \int \left(\varphi\frac{dE}{dx} \right)dx \tag{4-58}$$

which can be integrated to

$$W = \varphi \cdot \Delta E \tag{4-59}$$

Equation 4-59 can be differentiated with ΔE held constant to obtain

$$dW = \Delta E \cdot d\varphi \tag{4-60}$$

The power expended in displacing the charge dx is

$$P = \frac{dW}{dt} = \frac{(\Delta E \cdot d\varphi)}{dt} = \frac{d\varphi}{dt} \cdot \Delta E \tag{4-61}$$

The term $d\varphi/dt$ is equal to electrical current I. So Eq. 4-61 becomes

$$P = I \cdot \Delta E \tag{4-62}$$

Substituting P from Eq. 4-62 into $dW = P \cdot dt$ yields

$$dW = (I \cdot \Delta E) \cdot dt \tag{4-63}$$

Of course, energy may be transferred from one location to another, or from a control volume (system) to its surroundings, by mass transport. Energy is inherent in mass in the form of internal energy and is transported when the mass is transported.

These mechanisms of energy transfer must be accounted for in energy balances.

Energy Balances

The first law of thermodynamics is also known as the conservation of energy. It is a quantitative balance—an accounting of the various energy forms in the system and the means by which the energy is transferred. Simply stated, the first law of thermodynamics is as follows:

The total energy of a system and its surroundings is always conserved.

Although energy assumes many forms, the total quantity of energy is constant, and when energy disappears in one form it appears simultaneously in other forms (Smith and Van Ness 1987). Note that the first law of thermodynamics is for the system of interest *and* its surroundings:

$$\Delta(\text{energy of the system}) + \Delta(\text{energy of the surroundings}) = 0 \tag{4-64}$$

An energy balance can be performed on a *system* in a similar fashion to how balances are created in general, by accounting for all energy storage, transformation, and transfer. The following word equation describes the general energy balance:

$$\text{accumulation} = \text{flow in} - \text{flow out} + \text{creation} - \text{destruction} \tag{4-65}$$

Since energy cannot be created nor destroyed, the energy balance reduces to

$$\text{energy accumulation} = \text{energy flow in} - \text{energy flow out} \tag{4-66}$$

For the general system shown in Fig. 4-5, we can determine the quantity of energy accumulating in the system and the amount of energy flowing in and flowing out. Energy accumulation within the system is in the form of internal, potential, and kinetic energy:

$$d[m(u + pe + ke)]_{\text{sys}} \tag{4-67}$$

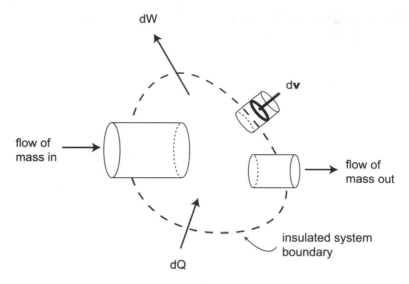

Figure 4-5. General system for energy balance showing energy transfers.

where m is the mass within the control volume, or system. The energy flowing into the system is

$$(u + pe + ke)_{\text{in}} \, dm_{\text{in}} \qquad (4\text{-}68)$$

and the energy flowing out of the system is

$$(u + pe + ke)_{\text{out}} \, dm_{\text{out}} \qquad (4\text{-}69)$$

Energy can also be transformed into heat and mechanical work:

dQ = the net heat transferred *to* the system and
dW = the net work done *by* the system.

Note the sign conventions: Positive dQ is net energy transferred to the system, and positive dW is net work done by the system.

Substituting these terms into the energy balance in Eq. 4-66 gives

$$
\begin{aligned}
d[m(u + pe + ke)]_{\text{sys}} = {}& (u + pe + ke)_{\text{in}} \, dm_{\text{in}} \\
& - (u + pe + ke)_{\text{out}} \, dm_{\text{out}} + dQ - dW
\end{aligned}
\qquad (4\text{-}70)
$$

The signs on dQ and dW, denoting the directions of energy flow for dQ and dW (heat transferred to and work done by) are by convention. The directions of energy flow for dQ and dW can be reversed (i.e., heat transferred from and work done to), in which case the signs for dQ and dW would be reversed in the energy balance.

There are several assumptions made in the derivation of this equation that must now be explicitly stated:

1. The properties within the system are homogeneous and uniform.
2. The properties in the inflow and outflow streams are homogeneous and uniform.
3. Gravitational acceleration is constant.

Although dW is usually considered to be for mechanical work, such as shaft work, or electrical work, the term can also be used for other forms of work from energy transfer (e.g., interfacial energy or magnetic energy).

Substituting the definitions of potential and kinetic energies into Eq. 4-70 gives an energy balance of the form

$$d\left[m\left(u + gz + \frac{V^2}{2}\right)\right]_{sys} = \left(u + gz + \frac{V^2}{2}\right)_{in} dm_{in}$$

$$-\left(u + gz + \frac{V^2}{2}\right)_{out} dm_{out} + dQ - dW \qquad (4\text{-}71)$$

For an open system where mass flows into and out of the system, work is done against pressure P:

$$dW = Fdx = (PA)dx = Pd\mathbf{v} \qquad (4\text{-}72)$$

The value of $d\mathbf{v}$ is a function of the specific volume, v, of the substance:

$$d\mathbf{v} = v \cdot dm \qquad (4\text{-}73)$$

For mass flowing into a system, from Eq. 4-72 and Eq. 4-73, we have

$$dW = P_{in}d\mathbf{v}_{in} = P_{in} v_{in} dm_{in} \qquad (4\text{-}74)$$

Similarly for mass flow out of a system, we have

$$dW = P_{out}d\mathbf{v}_{out} = P_{out} v_{out} dm_{out} \qquad (4\text{-}75)$$

This work is known as "injection work," flow work, or flow energy. It is the work that must be expended to inject a mass "packet" of dm across system boundaries.

Adding this additional work term to our energy balance we obtain

$$d\left[m\left(u + gz + \frac{V^2}{2}\right)\right]_{sys} = \left(u + Pv + gz + \frac{V^2}{2}\right)_{in} dm_{in}$$

$$-\left(u + Pv + gz + \frac{V^2}{2}\right)_{out} dm_{out} + dQ - dW \qquad (4\text{-}76)$$

The two terms u and Pv can be combined into one term called *enthalpy*. The specific enthalpy h, or enthalpy per unit mass, is defined as

$$h = u + Pv \tag{4-77}$$

By substituting this definition into Eq. 4-76, the energy balance becomes

$$d\left[m\left(u + gz + \frac{V^2}{2} \right) \right]_{sys} = \left(h + gz + \frac{V^2}{2} \right)_{in} dm_{in}$$

$$- \left(h + gz + \frac{V^2}{2} \right)_{out} dm_{out} + dQ - dW \tag{4-78}$$

This equation can be simplified for an open system at steady flow to

$$dh + gdz + d\left(\frac{V^2}{2} \right) = \frac{dQ}{dm} - \frac{dW}{dm} \tag{4-79}$$

To make our energy balance more general to account for more than one flow of mass into and out of the system, we can use a summation for the inflow and out-flow terms:

$$d\left[m\left(u + gz + \frac{V^2}{2} \right) \right]_{sys} = \sum_{in}\left[\left(h + gz + \frac{V^2}{2} \right)_{in} dm_{in} \right]$$

$$- \sum_{out}\left[\left(h + gz + \frac{V^2}{2} \right)_{out} dm_{out} \right] + dQ - dW \tag{4-80}$$

Summations can also be used for dQ and dW in the event that there is more than one path by which heat or work is transferred to or from the system.

Bernoulli's Equation

The energy balance derived in Eq. 4-78 accounts for all possible energy transformations and transfers in any system. However, this energy balance may be simplified for application to hydraulic systems. When the state of mass does not change with time (i.e., conditions and properties remain temporally constant), the system is considered to be *steady state*. For steady state and steady flow with only one mass inflow and one mass outflow, the energy balance may be condensed to

$$\left(u + Pv + gz + \frac{V^2}{2} \right)_{in} - \left(u + Pv + gz + \frac{V^2}{2} \right)_{out} = \frac{dW}{dm} - \frac{dQ}{dm} \tag{4-81}$$

where $dm = dm_{in} = dm_{out}$. This may be further simplified to

$$\Delta\left(Pv + gz + \frac{V^2}{2}\right) = \frac{-dW}{dm} - \left(\Delta u - \frac{dQ}{dm}\right) \tag{4-82}$$

The inverse of the specific volume, v, is density, ρ, so

$$\Delta\left(\frac{P}{\rho} + gz + \frac{V^2}{2}\right) = \frac{-dW}{dm} - \left(\Delta u - \frac{dQ}{dm}\right) \tag{4-83}$$

Friction heating per unit mass can be defined as

$$\Im = \Delta u - \frac{dQ}{dm} \tag{4-84}$$

This term accounts for friction "loss" or the "destruction" of energy within the system. Note that *energy is not lost or destroyed because of friction*, but it is converted to heat. The energy balance then becomes

$$\Delta\left(\frac{P}{\rho} + gz + \frac{V^2}{2}\right) = \frac{-dW}{dm} - \Im \tag{4-85}$$

This is *Bernoulli's equation*, an extremely important equation used for understanding hydraulic systems and all types of fluid flow. Because heat, dQ, and internal energy, u, are not explicitly contained in Bernoulli's equation, it is considered a mechanical energy balance. However, the heat and internal energy terms are grouped into the friction loss term, although it may not be apparent. Note that Bernoulli's equation assumes incompressible steady flow and is written from point to point along a fluid streamline.

A convenient form of Bernoulli's equation is the head form, obtained by dividing Eq. 4-85 by g:

$$\Delta\left(\frac{P}{\rho g} + z + \frac{V^2}{2g}\right) = \frac{-dW}{gdm} - \frac{\Im}{g} \tag{4-86}$$

Each term in the head form of Bernoulli's equation has units of length. The pressure form of Bernoulli's equation is

$$\Delta\left(P + \rho gz + \frac{\rho V^2}{2}\right) = \frac{-\rho dW}{dm} - \rho\Im \tag{4-87}$$

Each term in the pressure form of Bernoulli's equation has units of pressure.

Example (Adapted from Wilkes 1999)

A water storage tank (tank b), previously empty, is being filled from another tank (tank a) through a 4-in. inner diameter pipe as shown in Fig. 4-6. The water height in the full tank remains constant at $h_{\text{tank a}}$. The water height in the tank that is being filled (tank b) is h, and it increases until it is filled to its maximum height, $h_{\text{tank b}}$, which is the same height as tank a. Tank b is 25-ft wide by 25-ft long by 25-ft deep. Both tanks are open to the atmosphere. How long will it take for tank b to fill? Neglect friction.

Solution

Bernoulli's equation is

$$\Delta\left(\frac{P}{\rho g} + z + \frac{V^2}{2g}\right) = \frac{-dW}{gdm} - \frac{\Im}{g}$$

Writing Bernoulli's equation from point 1 to point 2, we have

$$\left(\frac{P}{\rho g} + z + \frac{V^2}{2g}\right)\Bigg|_2 - \left(\frac{P}{\rho g} + z + \frac{V^2}{2g}\right)\Bigg|_1 = 0$$

The pressure at point 1 is zero (atmospheric), and the velocity of the fluid at point 1 is zero. The static height, z, at point 2 is zero, and the pressure at point 2 is due to the head of water above it (h). The velocity of the water through the pipe is V. So Bernoulli's equation simplifies to

$$h_{\text{tank a}} - \left(h + \frac{V^2}{2g}\right) = 0$$

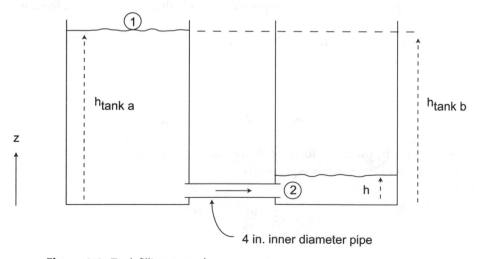

Figure 4-6. Tank filling example.

Solving for the velocity in the pipe, V, we get

$$V = \sqrt{2g(h_{\text{tank a}} - h)}$$

Performing a mass balance on tank b with Eq. 4-20 gives

$$(\dot{m}_{\text{in}} \cdot \Delta t) + 0 = (\dot{m}_{\text{out}} \cdot \Delta t) + \Delta m$$

There is no flow out, and the mass flow rate in is $\rho \cdot V \cdot A_{\text{pipe}}$, so

$$\rho V A_{\text{pipe}} = \frac{\Delta m}{\Delta t}$$

The change in mass, Δm, in tank b is $\rho \cdot A_{\text{tank b}} \cdot \Delta h$, where $A_{\text{tank b}}$ is the cross-sectional area (length $\times$ width) of the tank, so

$$\rho V A_{\text{pipe}} = \rho A_{\text{tank b}} \frac{\Delta h}{\Delta t}$$

Taking the limits $\Delta t \to 0$, $\Delta h \to 0$ yields the differential equation

$$\rho V A_{\text{pipe}} = \rho A_{\text{tank b}} \frac{dh}{dt}$$

Substituting $V = \sqrt{2g(h_{\text{tank a}} - h)}$ for V gives

$$\rho A_{\text{pipe}} \sqrt{2g(h_{\text{tank a}} - h)} = \rho A_{\text{tank b}} \frac{dh}{dt}$$

This equation can be rearranged to get

$$\frac{dh}{dt} = \frac{A_{\text{pipe}}}{A_{\text{tank b}}} \sqrt{2g(h_{\text{tank a}} - h)}$$

Performing a separation of variables, we have

$$\frac{A_{\text{pipe}}}{A_{\text{tank b}}} \int_0^t dt = \int_0^{h_{\text{tank b}}} \frac{dh}{\sqrt{2g(h_{\text{tank a}} - h)}}$$

Integration then yields

$$t \Big|_0^t \frac{A_{\text{tank b}}}{A_{\text{pipe}}} \left[-\frac{1}{g} \sqrt{2g(h_{\text{tank a}} - h)} \right]_0^{h_{\text{tank b}}}$$

Noting that $h_{\text{tank a}} = h_{\text{tank b}}$ we can solve this equation for time, obtaining

$$t = \frac{A_{\text{tank b}}}{A_{\text{pipe}}}\left[\sqrt{\frac{2(h_{\text{tank a}})}{g}}\right]$$

Substituting in numerical values then gives

$$t = \frac{25\text{ ft} \cdot 25\text{ ft}}{\dfrac{\pi}{d}\left(\dfrac{4}{12}\text{ ft}\right)^2}\cdot\left[\sqrt{\frac{2(25\text{ ft})}{32.2\dfrac{\text{ft}}{\text{s}^2}}}\right] = 8925\text{ s or }149\text{ min}$$

Symbol List

a	acceleration
A	cross-sectional area
E	potential, voltage
$\vec{f}$	force per unit volume
F	force
$\mathfrak{F}$	frictional loss
g	acceleration due to gravity
g_c	Newton's law proportionality factor, $32.174\text{ ft}\cdot\text{lbm/lbf}\cdot\text{s}^2$ for the U.S. Customary System of units
h	specific enthalpy
I	current
ke	specific kinetic energy
KE	kinetic energy
L	length
m	mass
$\dot{m}$	mass flow rate
P	power, pressure
pe	specific potential energy
PE	potential energy
Q	volumetric flow rate, heat transfer
t	time
T	torque
u	specific internal energy
U	internal energy
v	specific volume
V	velocity
$\mathbf{v}$	volume
W	work

μ	absolute viscosity
θ	angular displacement
ρ	fluid density
τ	shear stress
ω	rotational velocity
φ	electrical charge

Problems

1. A municipal wastewater treatment system can treat a maximum of 950 m³ per day of wastewater. During a rain event, 2,000 m³ of untreated wastewater had to be stored to avoid exceeding the maximum system flow rate. Now that the rain event is over, this stored wastewater must be added to the current existing system influent, which is flowing at a constant 660 m³ per day. Assume a constant influent flow rate and that the stored wastewater is pumped at a constant flow rate. How many hours must the stored wastewater be "bled" into the influent stream?

2. A drinking water supply tank is supplied at 500 gpm through a pipe from a reservoir (see Fig. 4-7). The reservoir water surface is 200 ft higher than the surface of the water in the water tank. A turbine is installed in the pipe as shown in the Fig. 4-7. The frictional head loss in the pipe is 2.0 ft and the temperature is 15 °C. What power is produced by the turbine?

3. A small tank is being filled with water through an 8-cm-diameter pipe from a large pressurized storage tank with the system shown in Fig. 4-8. Assume that the frictional head loss is 1.0 m and that the volume of water in the large storage tank does not fluctuate. The large storage tank is pressurized with an

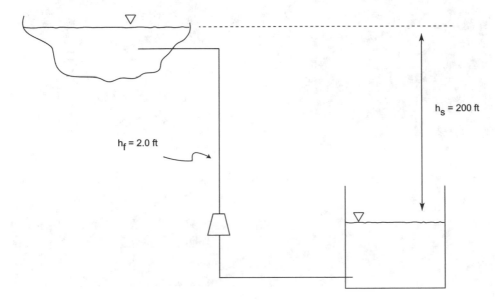

Figure 4-7. System in Problem 2.

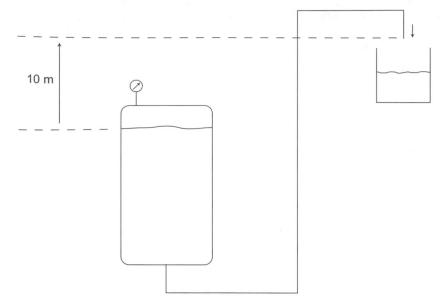

Figure 4-8. System in Problem 3.

external (pressurized) air supply. What air pressure must be in the large storage tank to supply a flow rate of 3.6 m³/hr through the pipe?

4. The distribution arms on a trickling filter distribute wastewater to media to allow biological activity to treat the waste (see Fig. 4-9). It is important that the

Figure 4-9. Trickling filter showing wastewater distribution arms (Problem 4).

wastewater be uniformly loaded over the surface of the trickling filter. This is usually done by varying orifice sizes for each port. Because the outer ports on the distribution arm must cover a greater surface area of the trickling filter per revolution than the inner ports, the outer ports must deliver greater flow rate. For a 20-ft-diameter trickling filter with ¼-in.-diameter discharge ports on the distribution arms, determine how much additional frictional loss the innermost port (located 1.5 ft from the centerline of the trickling filter, which is the pivot point of the distribution arm) must have in relation to the outermost port (located 9.5 ft from the centerline). Assume that each port distributes wastewater on a 1-ft-wide band on the surface of the filter and the flow rate through the innermost port is 2 gpm.

References

Lewis, G. N., and Randall, M. (1923). *Thermodynamics and the Free Energy of Chemical Substances*, McGraw-Hill, New York.

Munson, B. R., Young, D. F., and Okiishi, T. H. (2002). *Fundamentals of Fluid Mechanics*, Wiley, New York.

Smith, J. M., and Van Ness, H. C. (1987). *Introduction to Chemical Engineering Thermodynamics*, McGraw-Hill, New York.

Sposito, G. (1981). *The Thermodynamics of Soil Solutions*, Oxford University Press, New York.

Wilkes, J. O. (1999). *Fluid Mechanics for Chemical Engineers*, Prentice Hall, Upper Saddle River, NJ.

CHAPTER 5

Friction in Closed-Conduit Fluid Flow

Chapter Objectives

1. Identify laminar and turbulent fluid flow.
2. Quantify pressure losses in a closed-conduit hydraulic system with Newtonian flow.
3. Discuss boundary layers and transition length.
4. Calculate friction loss from valves and fittings.
5. Estimate friction factors from available correlations.
6. Apply flow and discharge coefficients.
7. Solve for pressure loss in non-Newtonian fluid flow.

In Chapter 4, Bernoulli's equation was derived from an energy balance:

$$\Delta\left(\frac{P}{\rho} + g \cdot z + \frac{V^2}{2}\right) = \frac{-dW}{dm} - \Im \qquad (5\text{-}1)$$

This equation shows the relationship among pressure P, fluid velocity V, elevation z, work W, and friction $\Im$, all of which may be considered energy terms. To be able to describe how a fluid system operates, the friction term in this equation must be quantified. Whereas it will be seen that many of the parameters that are used to calculate frictional losses for fluids flowing through conduits have been empirically determined, an understanding of the conceptual underpinnings of frictional loss and fluid flow is necessary to be able to competently analyze flow through closed conduits.

Fluid Flow Phenomena

Given a section of closed conduit as shown in Fig. 5-1, with a Newtonian fluid passing through the section at a uniform, controlled flow rate, the pressure drop versus flow rate behavior is found to be approximately described by Fig. 5-2. The pressure

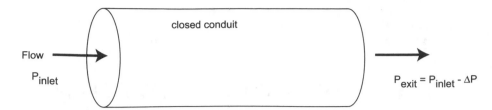

Figure 5-1. Flow through a closed conduit.

drop increases linearly as the flow rate is increased from zero, until a transition zone is reached. In the transition zone, the flow is somewhat unstable, and the pressure drop is not predictable with accuracy. Above the transition zone, the pressure drop is approximately a function of the flow rate squared.

In the late 1800s, through experiments using a dye, Osborne Reynolds was able to provide insight into the phenomena of fluid flow that produced the observations illustrated in Fig. 5-2. The dye showed the flow in the lower flow rate range to be without mixing—the injected dye ran straight and thin for the length of the pipe. In the high flow rate region, there was fast mixing of the dye. It was rapidly dispersed throughout the cross section of the pipe. These types of flow are now known as *laminar* and *turbulent*, respectively.

Laminar flow is characterized by flow without mixing. Layers of fluid adjacent to each other flow past one another without mixing. There are no eddies in laminar flow, just straight flow lines. Turbulent flow has chaotic motion in all directions, including lateral mixing. As flow rate is increased above that producing laminar flow, the flow becomes somewhat unstable with results that cannot be exactly reproduced or accurately predicted. This region of flow in between laminar and turbulent is called the *transition region* or *transition zone*.

A dimensionless number that characterizes fluid flow, the *Reynolds number*, has been found to delineate flow type. The Reynolds number $\Re$ is a dimensionless

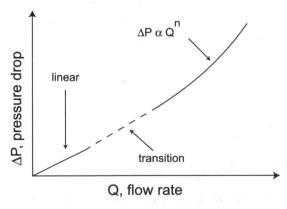

Figure 5-2. Experimental pressure-drop curve for the pipe section in Fig. 5-1. The value for *n* ranges from 1.8 to 2.0. *Source:* Adapted from Wilkes (1999).

group based on dynamic similitude, as it is a ratio of the inertial force to the viscous force. See Avallone and Baumeister (1996) for more information on dimensionless groups. The inertia force of a fluid element divided by the viscous force is

$$\Re \propto \frac{\text{inertial force}}{\text{viscous force}} = \frac{\text{mass} \cdot \text{acceleration}}{\text{viscous shear stress} \cdot \text{area}}$$

$$= \frac{(\rho \cdot L^3)\left(\dfrac{L}{T^2}\right)}{\tau \cdot L^2} = \frac{\rho \cdot L^2 \cdot V^2}{\mu \cdot \left(\dfrac{V}{L}\right) \cdot L^2} = \frac{\rho \cdot L^2 \cdot V^2}{\mu \cdot V \cdot L} \qquad (5\text{-}2)$$

where L and V are a characteristic length and characteristic velocity in the system, respectively, ρ is the fluid density, and μ is the fluid viscosity.

Therefore the Reynolds number is a function of a characteristic length, a characteristic velocity, the fluid density, and the fluid viscosity:

$$\Re = \text{func}\left(\frac{\rho \cdot L \cdot V}{\mu}\right) \qquad (5\text{-}3)$$

The characteristic length and velocity are specifically defined for pipe flow as the pipe diameter and average fluid velocity, respectively. These definitions are somewhat obvious, but the characteristic length and velocity may not be as obvious for other flow systems. For example, for flow through porous media (such as in sand filtration), the pore diameter or grain diameter could be used as a characteristic length. The fluid velocity could be the superficial fluid velocity or the fluid velocity in the pore. This is covered later in the text. So although it is quite intuitive as to what characteristic length and velocity are used in the Reynolds number for pipe flow, for other systems it is important to verify what length and velocity are used to calculate Reynolds number.

For closed-conduit, circular pipe of diameter D, the Reynolds number for Newtonian flow through pipes is calculated with the relationship

$$\Re = \frac{D \cdot V \cdot \rho}{\mu} \qquad (5\text{-}4)$$

Sometimes other symbols are used for Reynolds number, such as N_{Re} and Re.

For this definition of Reynolds number, laminar flow generally occurs up through a Reynolds number of 2,000 in pipe flow. Above 2,000, the transition zone first appears in pipe flow, and as the Reynolds number increases, the flow becomes turbulent. Ordinarily, above a Reynolds number of approximately 4,000, turbulent flow exists in pipes. However, under some conditions, laminar flow can persist at much higher Reynolds numbers.

Laminar Flow

To gain an understanding of laminar flow, consider a disk-shaped fluid element of incompressible Newtonian fluid flowing at steady state in a circular tube as in Fig. 5-3. Assume the fluid is flowing without lateral mixing (i.e., the flow is laminar).

The forces acting on the fluid element are pressure forces at each face of the disk and forces from fluid shear on the sides of the disk. The net pressure force, which is the sum of the pressure forces on each face of the disk element, is

$$F_{\text{pressure}} = P(\pi \cdot r^2) - (P - \Delta P)(\pi \cdot r^2) \tag{5-5}$$

where P is the pressure on the left disk face of disk (the upstream face), ΔP is the difference in pressure between the upstream face and downstream face, and r is the radius of the disk-shaped element.

The force on the fluid element resulting from shear stress is

$$F_{\text{shear}} = (2\pi \cdot r) \cdot \Delta x \cdot \tau \tag{5-6}$$

where Δx is the thickness of the disk over which the shear stress τ acts.

For the fluid element to remain static, the sum of the forces on the fluid element must be zero (by Newton's second law). The sum of the forces on this element is

$$F_{\text{pressure}} + F_{\text{shear}} = 0 \tag{5-7}$$

Using Eqs. 5-5 and 5-6 in Eq. 5-7 gives

$$[(\Delta P)(\pi \cdot r^2)] + [(2\pi \cdot r) \cdot \Delta x \cdot \tau] = 0 \tag{5-8}$$

Rearranging to solve for shear stress, we have

$$\tau = -\frac{r(\Delta P)}{2 \cdot \Delta x} \tag{5-9}$$

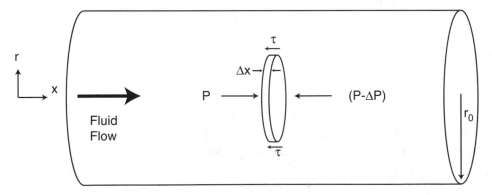

Figure 5-3. Disk-shaped element in steady fluid flow in a closed conduit.
Source: Adapted from McCabe et al. (1993).

Newton's law of viscosity is

$$\tau = \mu \frac{dV}{dr} \tag{5-10}$$

Setting Newton's law of viscosity, Eq. 5-10, equal to Eq. 5-9, eliminates τ and yields

$$\mu \frac{dV}{dr} = -\frac{r(\Delta P)}{2 \cdot \Delta x} \tag{5-11}$$

Separating variables then leads to

$$\mu dV = -\frac{(\Delta P)}{2 \cdot \Delta x} r dr \tag{5-12}$$

Assuming that $\Delta P/\Delta x$ does not vary with radius, and integrating from $r = r_0$ (the pipe wall), where $V = 0$ (the "no-slip" assumption), to $r = r$, where $V = V$, we get

$$\mu \int_0^V dV = -\frac{(\Delta P)}{2 \cdot \Delta x} \int_{r_0}^r r dr \tag{5-13}$$

$$\mu V = -\frac{(\Delta P)}{4 \cdot \Delta x} r^2 \Big|_{r_0}^r \tag{5-14}$$

$$V = -\frac{(\Delta P)}{4\mu \cdot \Delta x} (r^2 - r_0^2) \tag{5-15}$$

Rearranging gives

$$V = \frac{(r_0^2 - r^2)}{4\mu} \cdot \frac{(\Delta P)}{\Delta x} \tag{5-16}$$

A plot of this equation, which describes the velocity profile for laminar flow, is shown in Fig. 5-4.

From Fig. 5-4, it can be seen that (a) the fluid velocity is zero at the wall of the conduit, (b) the maximum velocity occurs at the centerline of the conduit, and (c) the velocity profile for laminar flow is parabolic. After rearranging Eq. 5-16 to solve for ΔP, it can also be seen that ΔP depends on velocity and fluid viscosity, and is independent of fluid density:

$$\Delta P \propto V \cdot \Delta x \cdot \mu \tag{5-17}$$

Solving Eq. 5-16 for the maximum velocity at $r = 0$ yields

$$V_{max} = \frac{r_0^2}{4\mu} \cdot \frac{\Delta P}{\Delta x} \tag{5-18}$$

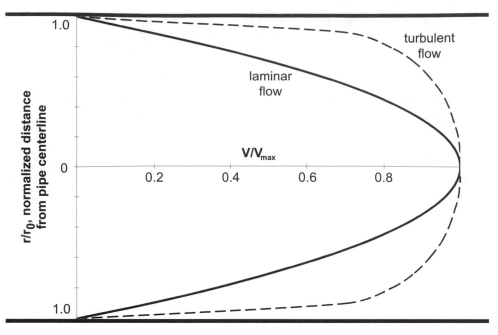

Figure 5-4. Velocity distribution in a closed conduit for laminar and turbulent flow. *Source:* Adapted from McCabe et al. (1993).

The average velocity can be found from

$$V = \frac{\int_0^{r_0} V \cdot r\,dr}{\int_0^{r_0} r\,dr} \tag{5-19}$$

Substituting Eq. 5-16 into Eq. 5-19 for V gives

$$V = \frac{\int_0^{r_0} \left(\dfrac{r_0^2 - r^2}{4\mu} \dfrac{\Delta P}{\Delta x} \right) \cdot r\,dr}{\int_0^{r_0} r\,dr} \tag{5-20}$$

Integrating the numerator and denominator, we have

$$V = \frac{\left(\dfrac{r_0^4}{16\mu} \dfrac{\Delta P}{\Delta x} \right)}{\left(\dfrac{r_0^2}{2} \right)} \tag{5-21}$$

which can be simplified to

$$V = \frac{r_0^2}{8\mu} \frac{\Delta P}{\Delta x}$$
(5-22)

Dividing Eq. 5-18 by Eq. 5-22 gives us the ratio

$$\frac{V_{max}}{V} = \frac{\dfrac{r_0^2}{4\mu} \cdot \dfrac{\Delta P}{\Delta x}}{\dfrac{r_0^2}{8\mu} \cdot \dfrac{\Delta P}{\Delta x}}$$
(5-23)

Canceling identical terms in the numerator and denominator then yields

$$\frac{V_{max}}{V} = 2$$
(5-24)

So the flow can be averaged across the radius to find an average fluid velocity in the conduit. This derivation shows that the maximum fluid velocity, which takes place at the centerline of the conduit, is twice the average fluid velocity in the conduit for laminar flow.

If a fluid flows through a closed conduit, the friction exerted on the fluid will result in a lower pressure in the conduit as the fluid travels along: The fluid loses energy. The pressure drop through a closed conduit is of great interest to engineers. Being able to predict the pressure loss, knowing values for governing parameters, is absolutely necessary for engineers designing and troubleshooting fluid systems. For laminar flow, Eq. 5-22 can be rearranged to solve for ΔP:

$$\Delta P = \frac{8 \cdot \mu \cdot \Delta x \cdot V}{r_0^2}$$
(5-25)

Since the diameter of the closed conduit, D_0, is twice the radius r_0 of the conduit, we have

$$\Delta P = \frac{8 \cdot \mu \cdot \Delta x \cdot V}{\dfrac{D_0^2}{4}} = \frac{32 \cdot \mu \cdot \Delta x \cdot V}{D_0^2}$$
(5-26)

The flow rate through a circular tube is a function of the average velocity V and the flow area:

$$Q = V \cdot \text{flow area}$$
(5-27)

Substituting an equation for the flow area gives

$$Q = V \cdot \left(\pi \frac{D_0^2}{4} \right) \qquad (5\text{-}28)$$

Rearranging to solve for V then gives

$$V = \frac{4 \cdot Q}{\pi \cdot D_0^2} \qquad (5\text{-}29)$$

Substituting Eq. 5-29 into Eq. 5-26 yields an equation explicit in ΔP:

$$\Delta P = \left(\frac{32 \cdot \mu \cdot \Delta x}{D_0^2} \right) \left(\frac{4 \cdot Q}{\pi \cdot D_0^2} \right) \qquad (5\text{-}30)$$

which simplifies to

$$\Delta P = \frac{128 \cdot \mu \cdot \Delta x \cdot Q}{\pi \cdot D_0^4} \qquad (5\text{-}31)$$

This is known as the *Hagen–Poiseuille* equation. It describes the frictional loss or pressure drop in a circular closed conduit with laminar flow, where Δx is the section length. So, if the viscosity of the fluid is known, and the diameter of the conduit is known, the pressure loss as a function of length and flow rate can be calculated. The Hagen–Poiseuille equation is very useful for engineers as it can predict pressure loss in a conduit for laminar flow.

Example

Consider a 300-m-long pipe carrying 20 °C (68 °F) water at 7.5 L/m. The inner diameter of the pipe is 15 cm. What is the pressure drop in pascals?

Solution

The flow rate is

$$Q = 7.5 \text{ L/m} = 1.25 \times 10^{-4} \text{ m}^3/\text{s}$$

and the viscosity of water at 20 °C is

$$\mu = 1.0 \text{ cP} = 1.0 \times 10^{-3} \frac{\text{kg}}{\text{m·s}}$$

The Hagen–Poiseuille equation is

$$\Delta P = \frac{128 \cdot \mu \cdot \Delta x \cdot Q}{\pi \cdot D_0^4}$$

Substituting in the numerical values gives us

$$\Delta P = \frac{128 \left(1.0 \times 10^{-3} \, \frac{kg}{m \cdot s}\right)(300 \, m)\left(1.25 \times 10^{-4} \, \frac{m^3}{s}\right)}{\pi (0.15 \, m)^4}$$

or

$$\Delta P = 3.0 \frac{kg}{m \cdot s^2}$$

Converting to units of pascals, we obtain

$$\Delta P = 3.0 \, Pa$$

The Hagen–Poiseulle equation only applies for laminar flow, $\Re < 2,000$, which needs to be verified.

The average fluid velocity in the pipe is the flow rate divided by the area:

$$V = \frac{Q}{A} = \frac{1.25 \times 10^{-4} \, \frac{m^3}{s}}{\frac{\pi}{4}(0.15 \, m)^2} = 7.1 \times 10^{-3} \, \frac{m}{s}$$

We calculate the Reynolds number from

$$\Re = \frac{D \cdot V \cdot \rho}{\mu}$$

and so

$$\Re = \frac{0.15 \, m \cdot 7.1 \times 10^{-3} \, \frac{m}{s} \cdot 998 \frac{kg}{m^3}}{1.0 \times 10^{-3} \, \frac{kg}{m \cdot s}} = 1,059$$

The Reynolds number is below 2,000, so the Hagen–Poiseulle equation applies.

Friction loss can be put in a different, more general form. Recall Bernoulli's equation,

$$\Delta\left(\frac{P}{\rho} + g \cdot z + \frac{V^2}{2}\right) = \frac{-dW}{dm} - \Im \tag{5-32}$$

where z is the static height of the fluid. Writing Bernoulli's equation across the Δx thickness of the disk in Fig. 5-3 yields

$$\left.\left(\frac{P}{\rho} + g \cdot z + \frac{V^2}{2}\right)\right|_{x+\Delta x} - \left.\left(\frac{P}{\rho} + g \cdot z + \frac{V^2}{2}\right)\right|_{x} = \frac{-dW}{dm} - \Im \tag{5-33}$$

There is no work done on the fluid or by the fluid in the disk, so dW/dm is zero and

$$\left.\left(\frac{P}{\rho} + g \cdot z + \frac{V^2}{2}\right)\right|_{x+\Delta x} - \left.\left(\frac{P}{\rho} + g \cdot z + \frac{V^2}{2}\right)\right|_{x} = -\Im \tag{5-34}$$

Since the static height z is the same at both ends of the disk, and the velocity is constant:

$$\left.\left(\frac{P}{\rho}\right)\right|_{x+\Delta x} - \left.\left(\frac{P}{\rho}\right)\right|_{x} = -\Im \tag{5-35}$$

Since ρ is constant, we have

$$\frac{1}{\rho}\left(\left.P\right|_{x+\Delta x} - \left.(P)\right|_{x}\right) = -\Im \tag{5-36}$$

which simplifies to

$$\Im = \frac{\Delta P}{\rho} \tag{5-37}$$

Substituting in the Hagen–Poiseuille equation for ΔP yields an expression where the frictional loss $\Im$ can be calculated from known parameters of the system and fluid:

$$\Im = \frac{\left(\dfrac{128 \cdot \mu \cdot \Delta x \cdot Q}{\pi \cdot D_0^4}\right)}{\rho} = Q \cdot \Delta x \frac{\mu}{\rho} \frac{128}{\pi \cdot D_0^4} \tag{5-38}$$

We may define a new, dimensionless term, called the friction factor f, as

$$f = \frac{\Im}{4\left(\dfrac{\Delta x}{D}\right)\left(\dfrac{V^2}{2}\right)} \tag{5-39}$$

where $V = V_{ave}$ (the average fluid velocity) and $D = D_0$ (the pipe diameter). Substituting Eq. 5-38 for $\Im$ into Eq. 5-39 gives

$$f = \frac{\left(Q \cdot \Delta x \dfrac{\mu}{\rho} \dfrac{128}{\pi \cdot D^4} \right)}{4 \left(\dfrac{\Delta x}{D} \right) \left(\dfrac{V^2}{2} \right)} = \frac{\left[\left(V \dfrac{\pi \cdot D^2}{4} \right) \Delta x \dfrac{\mu}{\rho} \dfrac{128}{\pi \cdot D^4} \right]}{4 \left(\dfrac{\Delta x}{D} \right) \left(\dfrac{V^2}{2} \right)} \qquad (5\text{-}40)$$

Canceling terms in the numerator and denominator and substituting in the definition for Reynolds number yields for laminar flow

$$f = \frac{16}{\Re} \qquad (5\text{-}41)$$

Rearranging the definition for the friction factor (Eq. 5-39) to solve for $\Im$, we get

$$\Im = 4f \left(\frac{\Delta x}{D} \right) \left(\frac{V^2}{2} \right) \qquad (5\text{-}42)$$

We can now combine Eqs. 5-37 and 5-42 to form

$$\Im = \frac{\Delta P}{\rho} = 4f \left(\frac{\Delta x}{D} \right) \left(\frac{V^2}{2} \right) \qquad (5\text{-}43)$$

From this equation, we can calculate the friction loss term in Bernoulli's equation for laminar Newtonian flow and the pressure loss for a fluid flowing in a closed conduit, calculating the friction factor with Eq. 5-41.

Turbulent Flow

As discovered by Osborne Reynolds, when flow rate is increased through a conduit, transverse mixing occurs in the pipe above a certain flow rate—the flow becomes turbulent above a certain flow rate in a given pipe under constant conditions. In turbulent flow, eddies of various sizes exist to provide for transverse mixing of the fluid. The larger sized eddies are continuously formed in the flow stream and are broken down to smaller and smaller sizes. The small eddies dissipate energy as they finally disappear. In turbulence, energy is transferred from the bulk flow to the large eddies, to the smaller eddies, and finally to heat energy through viscous shear.

Because of the continuous formation and disappearance of eddies in turbulent flow, there is great variability in local, instantaneous fluid velocity, as shown in Fig. 5-5 (see McCabe et al. 1993).

Figure 5-5. Velocity fluctuations in turbulent flow. *Source:* Adapted from McCabe et al. (1993).

Because of the random nature of the eddy formation and mixing that takes place in turbulent flow, accurate quantitative determination of the flow profile and friction loss for turbulent flow cannot be accomplished a priori as for laminar flow. However, it has been found that the flow profile has a somewhat more flattened shape than laminar flow, as shown in Fig. 5-4. Turbulent flow is more "pluglike" in flow profile than laminar.

The friction loss for turbulent flow can be approached in a similar fashion as for laminar flow. Recall that for laminar flow the following we derived

$$\Im = \frac{\Delta P}{\rho} = 4f\left(\frac{\Delta x}{D}\right)\left(\frac{V^2}{2}\right) \tag{5-43}$$

For laminar flow we were able to calculate f from the Reynolds number with an equation derived from first principles. We cannot use first principles to derive an equation like this for turbulent flow, but values of f have been found as a function of Reynolds number and conduit surface roughness ε, as shown in Fig. 5-6. The value for the friction factor f can be used in Eq. 5-43 to calculate friction loss with turbulent flow.

Pipe roughness is characterized with a roughness length scale, ε. The roughness of a pipe is a characteristic of the pipe material and other factors such as degree of corrosion and amount of deposits. Typical surface roughnesses are

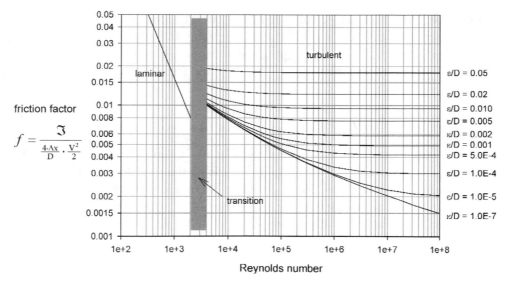

Figure 5-6. Chart of pipe friction factors as a function of relative roughness (ε/D) and Reynolds number.

listed in Table 5-1. The surface roughness ε can be expressed relative to the pipe diameter D, giving a *relative roughness*:

$$\text{relative roughness} = \frac{\varepsilon}{D} \tag{5-44}$$

By calculating the Reynolds number for fluid flow through a closed circular conduit, and through the use of Fig. 5-6 (by knowing the pipe wall roughness), a friction factor can be found. From our definition of friction factor, we can determine the friction loss of the flowing fluid with Eq. 5-33 and Fig. 5-6, for both laminar and turbulent flow.

Table 5-1. Typical surface roughnesses (clean pipe).

Material	$\varepsilon \; [mm]$	$\varepsilon \; [in.]$
Asphalt-dipped cast iron	0.12	4.8×10^{-3}
Cast iron	0.25	1.0×10^{-2}
Commercial steel	4.6×10^{-2}	1.8×10^{-3}
Concrete	0.3–3.0	1.2×10^{-2}–0.12
Drawn tubing	1.5×10^{-3}	6.0×10^{-5}
Galvanized iron	0.15	6.0×10^{-3}
Glass	1.5×10^{-3}	6.0×10^{-5}
Lead	1.5×10^{-3}	6.0×10^{-5}
Wood stave	0.18–0.91	7.2×10^{-3}–3.6×10^{-2}

Source: Data taken from Hydraulic Institute (1990) and Wilkes (1999).

Calculation of Friction Factors for Turbulent Flow

Determining the friction loss in a piping system requires knowledge of the friction factor. The friction factor for laminar flow is easily calculated from the equation derived earlier,

$$f = \frac{16}{\Re} \tag{5-41}$$

However, it is not quite as easy to *calculate* the friction factor for turbulent flow. The chart of friction factor versus Reynolds number can be used to find friction factors for turbulent flow (Fig. 5-6). Various empirical equations can be used as well. Some common equations for calculating f for turbulent flow, with their applicable ranges, are the following (McCabe et al. 1993):

$$f = 0.046 \cdot \Re^{-0.2} \text{ for } 5 \times 10^4 < R < 10^6, \text{smooth pipe} \tag{5-45}$$

$$f = 0.0014 + \frac{0.125}{\Re^{0.32}} \text{ for } 3 \times 10^3 < R < 3 \times 10^6, \text{smooth pipe} \tag{5-46}$$

and the Blasius equation,

$$f = 0.0790 \cdot \Re^{-0.25} \text{ for } R < 10^6, \text{smooth pipe} \tag{5-47}$$

However, these equations can only provide f for smooth pipe. None of them has any parameter or mechanism to adjust for ε, the roughness of the pipe. A more universally accepted relationship to calculate f for turbulent Newtonian flow is the Colebrook equation, which can take into account roughness:

$$\frac{1}{\sqrt{f}} = -4 \cdot \log \left(\frac{\frac{\varepsilon}{D}}{3.7} + \frac{1.255}{\Re\sqrt{f}} \right) \tag{5-48}$$

The Colebrook equation can handle pipes of varying roughness, but it is not explicit in f. The friction factor is on both sides of the equation, and so the solution must be found by trial and error or through root-finding software. One can use an iterative procedure, where a first estimate is made for f, that estimate is then substituted into the right-hand side of the equation, and the value for the left-hand side is solved for. The value of f that is found from this first iteration is then inserted back into the right-hand side, and the left-hand side is recalculated. It typically does not take many iterations for the solution to converge.

Shacham argued that one iteration with the Colebrook equation, starting with $f = 0.0075$ as the first estimate, converges to within 99% of the ultimate solution (Wilkes 1999). So an approximation of the Colebrook equation that is explicit in f, based on this assumption, is

$$f = \left\{ -1.737 \ln \left[0.269 \frac{\varepsilon}{D} - \frac{2.185}{\Re} \ln \left(0.269 \frac{\varepsilon}{D} + \frac{14.5}{\Re} \right) \right] \right\}^{-2} \tag{5-49}$$

Another empirical equation for finding the friction factor for Newtonian turbulent flow is Wood's approximation:

$$f = a + b\mathfrak{R}^{-c} \tag{5-50}$$

where

$$a = 0.0235\left(\frac{\varepsilon}{D}\right)^{0.225} + 0.1325\left(\frac{\varepsilon}{D}\right) \tag{5-51}$$

$$b = 22\left(\frac{\varepsilon}{D}\right)^{0.44} \tag{5-52}$$

$$c = 1.62\left(\frac{\varepsilon}{D}\right)^{0.134} \tag{5-53}$$

Example

Water at 50 °F (10 °C) flows at a flow rate of 625 gpm through a concrete pipe with an inner diameter of 1 ft and a length of 600 ft. What is the pressure drop in pounds per square inch?

Solution

The roughness ε of concrete pipe can be approximated as 5×10^{-3} ft, in the middle of the range of ε listed in Table 5-1. So the relative roughness is

$$\text{relative roughness} = \frac{\varepsilon}{D} = \frac{5.0 \times 10^{-3} \text{ ft}}{1 \text{ ft}} = 5.0 \times 10^{-3}$$

Converting the units of flow rate, we have

$$625 \text{ gpm} = 625 \frac{\text{gal}}{\text{min}} \cdot \frac{1 \text{ ft}^3}{7.48 \text{ gal}} \cdot \frac{1 \text{ min}}{60 \text{ s}} = 1.39 \frac{\text{ft}^3}{\text{s}}$$

The fluid velocity is

$$V = \frac{Q}{A} = \frac{1.39 \dfrac{\text{ft}^3}{\text{s}}}{\dfrac{\pi}{4}(1 \text{ ft})^2} = 1.77 \frac{\text{ft}}{\text{s}}$$

At 10 °C, the viscosity is

$$\mu = 1.35 \text{ cP} = 1.35 \text{ cP} \cdot \frac{6.72 \times 10^{-2} \dfrac{\text{lbm}}{\text{ft·s}}}{100 \text{ cP}} = 9.07 \times 10^{-4} \frac{\text{lbm}}{\text{ft·s}}$$

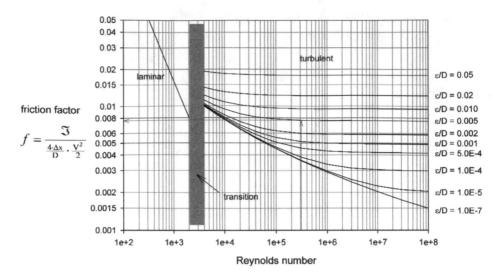

Figure 5-7. Pipe friction factor example.

and so the Reynolds number is

$$\Re = \frac{D \cdot V \cdot \rho}{\mu} = \frac{(1\,\text{ft}) \cdot 1.77\dfrac{\text{ft}}{\text{s}} \cdot 62.3\dfrac{\text{lbm}}{\text{ft}^3}}{9.07 \times 10^{-4}\dfrac{\text{lbm}}{\text{ft} \cdot \text{s}}} = 1.22 \times 10^5$$

For $\varepsilon/D = 5.0 \times 10^{-3}$ and $\Re = 1.22 \times 10^5$, f can be determined from the pipe friction factor chart (Fig. 5-6), which has been annotated in Fig. 5-7 for this example. From the figure, f is approximately 0.0075.

Rearranging Eq. 5-43 to solve for ΔP gives

$$\Delta P = \rho \cdot 4f \left(\frac{\Delta x}{D} \right) \left(\frac{V^2}{2} \right)$$

Substituting numerical values, we get

$$\Delta P = \frac{62.3\dfrac{\text{lbm}}{\text{ft}^3}}{32.174\dfrac{\text{ft} \cdot \text{lbm}}{\text{lbf} \cdot \text{s}^2}} \cdot 4(0.0075) \left(\frac{600\,\text{ft}}{1\,\text{ft}} \right) \left(\frac{\left(1.77\dfrac{\text{ft}}{\text{s}} \right)^2}{2} \right) = 54.6\dfrac{\text{lbf}}{\text{ft}^2}$$

or

$$\Delta P = 54.6\dfrac{\text{lbf}}{\text{ft}^2} \cdot \left(\frac{1\,\text{ft}}{12\,\text{in.}} \right)^2 = 0.38\,\text{psi}$$

Boundary Layers and Transition Length

When fluid flow is influenced by the presence of a solid boundary, it can form a layer composed of "sheets" of increasing lower velocities. Within this *boundary layer*, the velocity in the outermost layer is the maximum, whereas the velocity at the surface is zero, relative to the surface. For example, flow of fluid near a flat plate can form a boundary layer as shown in Fig. 5-8. The velocity of the fluid, u_{bulk}, at the upstream edge (the leading edge) is uniform. The fluid velocity at the fluid/solid interface is zero (the "no-slip" condition). As the fluid flows across the plate, a boundary layer forms, starting at the leading edge; this layer gets thicker, increasing Δz, with distance along the plate (in the x direction). Line BL separates the portion of the flow with a velocity gradient ($|\partial u/\partial z| > 0$) from that without a velocity gradient ($\partial u/\partial z = 0$). The portion of flow between the dashed line (BL) and the solid surface is called the *boundary layer*.

Flow in the boundary layer is at lower velocity than the bulk flow much further from the surface. Laminar flow with its straight flow lines and no mixing predominates near the surface, but eddies from the bulk turbulent flow occasionally "burst" into the boundary layer. However, because of their infrequent occurrence, these occasional eddies do not significantly affect the average velocities in the boundary layer.

At higher bulk velocities, turbulent flow can exist in the outer zone of the boundary layer. Between the turbulent zone of the boundary layer and the solid boundary, a viscous sublayer is present, as shown in Fig. 5-9.

Boundary layers form during fluid flow in closed conduits also. As fluid enters a straight tube, a boundary layer forms, starting at the entrance to the tube as illustrated in Fig. 5-10, as it does near a flat plate. Further from the entrance of the tube, the boundary layer thickens. Eventually, given enough straight length of tube, the boundary layer reaches the centerline of the tube. The velocity distribution in the tube is then unchanged with further distance along the tube, and the flow is considered *fully developed flow*.

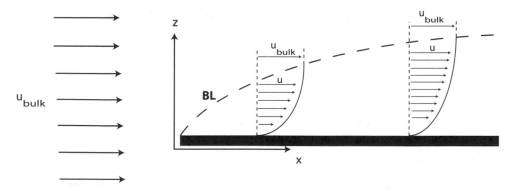

Figure 5-8. Boundary layer on a flat plate. *Source:* Adapted from McCabe et al. (1993).

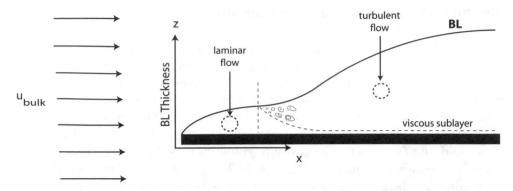

Figure 5-9. Zones of a typical boundary layer (laminar, turbulent and viscous sublayer) on a flat plate. *Source:* Adapted from McCabe et al. (1993).

Why is this important? The techniques for calculating the friction coefficient in pipes and tubes *assume fully developed flow*. The friction loss for fluid flow is dependent on the flow profile in the tube and therefore dependent on whether fully developed flow exists. Moreover, the flow profile is crucial for most instrumentation, as discussed in Chapter 9.

The tube length necessary for fully developed flow is called the *transition length*. For laminar flow the length of straight tube for fully developed flow can be calculated from (McCabe et al. 1993)

$$\frac{x_t}{D} = 0.05 \cdot \Re \quad \text{(laminar)} \tag{5-54}$$

For turbulent flow, the transition length is independent of Reynolds number. Approximately 40 to 50 pipe diameters are necessary to obtain fully developed flow for turbulent flow:

$$\frac{x_t}{D} = 40 \text{ to } 50 \quad \text{(turbulent)} \tag{5-55}$$

although 25 pipe diameters provides very close to fully developed flow.

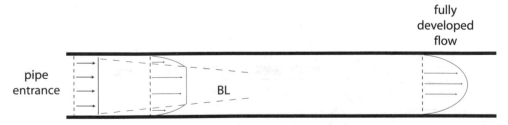

Figure 5-10. Boundary layer formation in closed conduits. *Source:* Adapted from McCabe et al. (1993).

Friction Reduction with Polymer Addition

Polymers in aqueous solution can produce a reduction in friction loss in flow through pipes. This was first noted by Toms in England in the 1940s (Levi 1995). Low concentrations of high-molecular-weight polymers such as polyethylene oxide and polyacrylamide can reduce pipe friction loss by up to 70%. Although the friction-reducing phenomenon has been observed, the mechanism causing the reduction in drag is only hypothesized at this point. It was found that delivery of the polymer near the pipe wall will bring about the reduction in flowing friction (Levi 1995). That is, the polymer does not have to be well dispersed in the water, just present near the pipe surface. It is thought that the polymers dissipate the small eddies that exist in the boundary layer, producing an increased viscous sublayer thickness. The increased thickness of the viscous sublayer is manifested in a smaller velocity gradient and thus a smaller wall shear (as expected from Newton's law of viscosity). Lower shear produces a reduced pressure drop in the flowing fluid. A low concentration of polyethylene oxide, on the order of parts per million, can significantly increase the flow rate through a pipe, given identical conditions. The effect of polyethylene oxide on friction reduction is shown in Fig. 5-11.

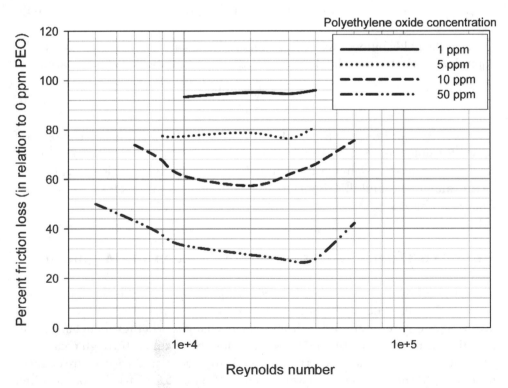

Figure 5-11. Friction reduction for dilute solutions of polyethylene oxide.
Source: Adapted from McCabe et al. (1993).

Flow through Noncircular Cross Sections

To estimate the friction in noncircular cross sections, the equivalent diameter is used in calculating the Reynolds number. The equivalent diameter D_{eq} is equal to four times the hydraulic radius r_H of the conduit,

$$D_{eq} = 4 \cdot r_H \tag{5-56}$$

where the hydraulic radius is defined as the ratio of the cross-sectional area S to the wetted perimeter L_p:

$$r_H = \frac{S}{L_p} \tag{5-57}$$

For a circular conduit completely full of fluid, the hydraulic radius is

$$r_H = \frac{S}{L_p} = \frac{\pi \dfrac{D^2}{4}}{\pi D} = \frac{D}{4} \tag{5-58}$$

so the equivalent diameter for a full pipe is

$$D_{eq} = 4 \cdot r_H = 4 \cdot \left(\frac{D}{4}\right) = D \tag{5-59}$$

The Reynolds number for noncircular section is calculated by substituting the equivalent diameter into the definition of Reynolds number presented earlier:

$$\Re = \frac{D_{eq} \cdot V \cdot \rho}{\mu} \tag{5-60}$$

For turbulent flow, the friction factor can be found from the correlations of f versus Reynolds number and surface roughness (ε) used for circular sections, as shown in Fig. 5-6. For laminar flow through noncircular cross sections, the friction factor cannot be found this simply; correlations exist in various texts for finding f from $\Re$ in this case.

Friction Loss from Changes in Velocity Direction and Magnitude

The friction loss that is accounted for with the friction factor f is that from "skin" friction of the fluid flowing near the solid surface, the wall of the conduit. Friction loss in piping systems also arises from velocity changes, whether that change is in direction or magnitude. When fluid streamlines are modified through changes in the conduit, the boundary layers are changed, and eddies and vortices are created, producing "form" friction, as shown in Fig. 5-12. This "form" friction occurs when fluid flows through valves and fittings used in piping systems and the fluid velocity direction or magnitude changes.

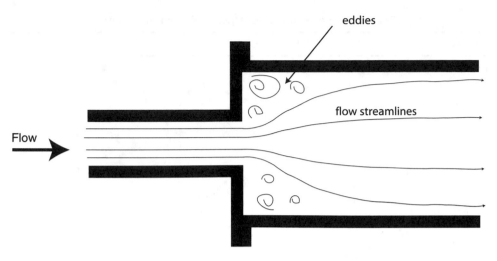

Figure 5-12. Flow disruption through sudden enlargement.

The actual flow profiles and friction losses that occur as the result of velocity changes cannot be determined accurately a priori, although computer simulations can be conducted. Fluid system designers have to rely on tabulated empirical data to calculate friction losses from velocity variations in fittings.

We derived the "skin" friction loss for fluids flowing through a pipe to be

$$\Im = 4f\left(\frac{\Delta x}{D}\right)\left(\frac{V^2}{2}\right) \tag{5-61}$$

The "form" friction loss from flow through fittings has been found to be a function of the average fluid velocity in the closed conduit (as it is for "skin" friction) and a resistance coefficient, K. The friction resulting from flow follows the relationship

$$\Im = K\left(\frac{V^2}{2}\right) \tag{5-62}$$

The resistance coefficients for fittings are found empirically for a specific fitting of a given design and manufacturer, and these values should be taken as an estimate. They are listed in many sources, including Crane's technical paper (Crane Co. 1985), the Hydraulic Institute's manual (Hydraulic Institute 1961), and others. Resistance coefficients (K values) for various fittings and valves are shown in Table 5-2 and Figs. 5-13, 5-14, and 5-15.

For quantifying friction losses through valves and fittings, resistance coefficients, K values, can be used if available for the specific component. However, there are alternate ways in which friction losses for valves and fittings are calculated. Flow coefficients are frequently provided by valve manufacturers, and orifices may be characterized with discharge coefficients.

Table 5-2. Typical fitting resistance coefficients (*K* values).

Fitting	K
Bell-mouth inlet	0.05
Square-edged inlet	0.5
Inward projecting inlet	1.0
Discharge	1.0
90° miter	1.129
45° miter	0.236
30° miter	0.130

Source: Data from Hydraulic Institute (1961).

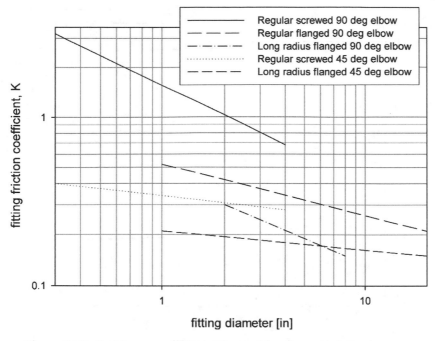

Figure 5-13. Resistance coefficients (*K* values) for elbows. Variation in resistance coefficients for elbows may be up to ±40%. *Source:* Data from Hydraulic Institute (1961).

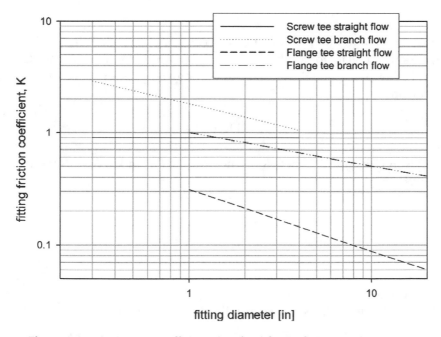

Figure 5-14. Resistance coefficients (*K* values) for tee fittings. Variation in resistance coefficients for tee fittings may be up to ±35%. *Source:* Data from Hydraulic Institute (1961).

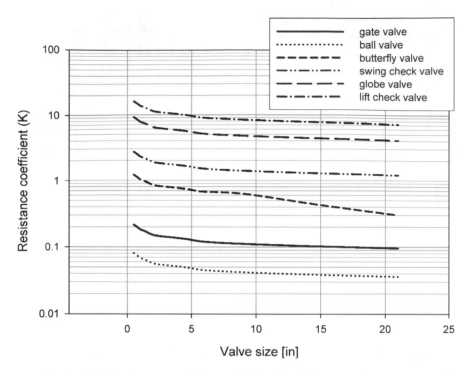

Figure 5-15. Resistance coefficients (*K* values) for fully open valves.
Source: Data from Crane Co. (1985).

It is common to find the flow characteristics of valves quantified with the flow coefficient, C_V. The C_V for a valve is determined by the flow rate of 15.6 °C (60 °F) water that can pass through the valve under a given pressure loss. The C_V is calculated from

$$C_V = 11.76 \frac{Q}{\sqrt{\Delta P}} \qquad (5\text{-}63)$$

when Q is in units of cubic meters per hour and ΔP is in kilopascals, and

$$C_V = \frac{Q}{\sqrt{\Delta P}} \qquad (5\text{-}64)$$

when Q is in units of gallons per minute and ΔP is in pounds per square inch.

Do not confuse C_V as strictly and directly representing the maximum flow capacity of the valve. It is just a convenient way to present the friction loss of that valve or fitting.

The friction loss across a valve installed in a system may be directly calculated from

$$\Delta P = \frac{11.76^2 \cdot Q^2 \cdot SG}{C_V^2} \qquad (5\text{-}65)$$

where SG is specific gravity of the liquid relative to 15.6 °C (60 °F) water, Q is in units of cubic meters per hour, and ΔP is in kilopascals, and

$$\Delta P = \frac{Q^2 \cdot SG}{C_V^2} \qquad (5\text{-}66)$$

when Q is in units of gallons per minute and ΔP is in pounds per square inch.

Example

Water at 10 °C flows at a flow rate of 2.5 m³/hr through a 40-mm-diameter valve with $C_V = 11.0$ (provided by the valve manufacturer). What is the pressure loss through the valve?

Solution

The density of water at 10 °C is 999.70 kg/m³ and at 15.6 °C it is 999.02 mg/m³, so the specific gravity is 999.70/999.02 = 1.001.

The pressure drop across the valve is calculated with

$$\Delta P = \frac{11.76^2 \cdot Q^2 \cdot SG}{C_V^2} = \frac{11.76^2 \cdot \left(2.5\,\frac{m^3}{hr}\right)^2 \cdot 1.001}{11.0^2} = 7.15\,\text{kPa}$$

Valve C_V can be converted to a resistance coefficient K, by using (Crane Co. 1985)

$$K = \frac{2.14 \times 10^{-3} \cdot d^4}{(C_V)^2} \quad \text{with } d \text{ in millimeters} \qquad (5\text{-}67)$$

or

$$K = \frac{891 \cdot d^4}{(C_V)^2} \quad \text{with } d \text{ in inches} \qquad (5\text{-}68)$$

where d is the inner diameter of the valve.

In a piping run consisting of valves, fittings, and pipe, the friction loss from fluid flow is the sum of losses from all components. So as the fluid flows through every pipe section, it loses energy because of friction; as the fluids passes through each valve, it loses energy; and as the fluid passes through each fitting (tee, elbow, etc.), it loses energy. To determine the total energy loss in a piping section, the frictional losses are simply summed:

$$\Im = \sum_{\substack{\text{all pipe} \\ \text{sections}}} \left\{ 4f\left(\frac{\Delta x}{D}\right)\left(\frac{V^2}{2}\right) \right\} + \sum_{\substack{\text{all valves} \\ \text{and fittings}}} \left\{ K\left(\frac{V^2}{2}\right) \right\} \qquad (5\text{-}69)$$

For a piping run of constant diameter, this simplifies to

$$\mathfrak{I} = 4f\left(\frac{\Delta x_{\text{total}}}{D}\right)\left(\frac{V^2}{2}\right) + \left(\sum_{\substack{\text{all valves} \\ \text{and fittings}}} K\right)\left(\frac{V^2}{2}\right) \tag{5-70}$$

When the fitting inner diameters are equal to the inner diameter of the pipe, the average fluid velocities are equal and so

$$\mathfrak{I} = \left(4\frac{f\Delta x_{\text{total}}}{D} + \sum_{\substack{\text{all valves} \\ \text{and fittings}}} K\right)\left(\frac{V^2}{2}\right) \tag{5-71}$$

Since $\mathfrak{I} = \delta P/\rho$, the total pressure loss is calculated with

$$\Delta P = \rho\left(4\frac{f\Delta x_{\text{total}}}{D} + \sum_{\substack{\text{all valves} \\ \text{and fittings}}} K\right)\left(\frac{V^2}{2}\right) \tag{5-72}$$

The head form, where the resultant friction loss h_f is reported in head units (e.g., feet, meters), is

$$h_f = \left(4\frac{f\Delta x_{\text{total}}}{D} + \sum_{\substack{\text{all valves} \\ \text{and fittings}}} K\right)\left(\frac{V^2}{2g}\right) \tag{5-73}$$

It is important to understand that summing the individual loss coefficients in a piping run to determine overall friction loss is an approximation. The flow profile and velocity distribution of a fluid are altered as the fluid passes through each component in a system, and the flow characteristics entering each component can affect the friction loss of that specific component. Figure 5-16 shows two computer simulation results—(a) for a 90° elbow at the end of a straight pipe and (b) for an elbow preceded by a gate representing a gate valve. The simulations were of the two-dimensional Navier–Stokes equations (see Chapter 4). It can be seen that the flow profile through the elbow is different with the presence of the gate upstream. It should be expected that the friction loss may be impacted by this change in flow profile.

Example

A 2-in. inner diameter piping run consists of one fully open globe valve ($C_V = 23$), four fully open gate valves ($C_V = 118$), five elbows, and six separate pipe lengths totaling 90 ft in length. Water at 68 °F (20 °C) flows through the piping run at 100 gpm. The piping is drawn tubing steel. What is the head loss in feet?

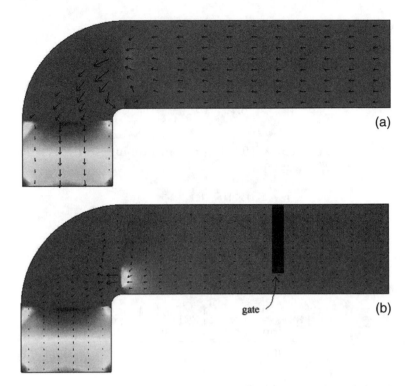

Figure 5-16. Two-dimensional Navier–Stokes simulation results for (a) flow through an elbow following a straight pipe and (b) flow through an elbow that follows a gate representing a gate valve. The length of the arrows is proportional to local velocity and the shading represents pressure (with darker shading indicating greater pressure).

Solution

The flow rate is

$$100 \, \text{gpm} = 100 \frac{\text{gal}}{\text{min}} \cdot \frac{1 \, \text{ft}^3}{7.48 \, \text{gal}} \cdot \frac{1 \, \text{min}}{60 \, \text{s}} = 0.223 \frac{\text{ft}^3}{\text{s}}$$

The fluid velocity is

$$V = \frac{Q}{A} = \frac{0.223 \frac{\text{ft}^3}{\text{s}}}{\frac{\pi}{4} \left(\frac{2}{12} \text{ft} \right)^2} = 10.2 \frac{\text{ft}}{\text{s}}$$

At 20 °C,

$$\mu = 1.0 \, \text{cP} = 1.0 \, \text{cP} \cdot \frac{6.72 \times 10^{-2} \frac{\text{lbm}}{\text{ft} \cdot \text{s}}}{100 \, \text{cP}} = 6.72 \times 10^{-4} \frac{\text{lbm}}{\text{ft} \cdot \text{s}}$$

and so the Reynolds number is

$$\Re = \frac{D \cdot V \cdot \rho}{\mu} = \frac{\left(\dfrac{2}{12}\,\text{ft}\right) \cdot 10.2\,\dfrac{\text{ft}}{\text{s}} \cdot 62.3\,\dfrac{\text{lbm}}{\text{ft}^3}}{6.72 \times 10^{-4}\,\dfrac{\text{lbm}}{\text{f} \cdot \text{s}}} = 1.58 \times 10^5$$

From Table 5-1, $\varepsilon = 6.0 \times 10^{-5}$ ft for drawn tubing, so the relative roughness is

$$\frac{\varepsilon}{D} = \frac{6.0 \times 10^{-5}\,\text{ft}}{2\,\text{in.}} = 3.0 \times 10^{-5}$$

From Fig. 5-6, $f = 0.004$.
For the globe valve $C_V = 23$ and the resistance coefficient is

$$K_{\text{globe}} = \frac{\left(891 \cdot \dfrac{1}{\text{in.}^4}\right) \cdot d^4}{(C_V)^2} = \frac{\left(891 \cdot \dfrac{1}{\text{in.}^4}\right) \cdot (2\,\text{in.})^4}{(23^2)} = 27$$

For the gate valves $C_V = 118$ and

$$K_{\text{gate}} = \frac{\left(891 \cdot \dfrac{1}{\text{in.}^4}\right) \cdot d^4}{(C_V)^2} = \frac{\left(891 \cdot \dfrac{1}{\text{in.}^4}\right) \cdot (2\,\text{in.})^4}{(118^2)} = 1.0$$

From Fig. 5-12 for a regular flanged 90° elbow, the elbows have $K = 0.42$. The sum of the resistance coefficients is

$$\sum_{\substack{\text{all valves} \\ \text{and fittings}}} K = 27 + (4 \cdot 1.0) + (5 \cdot 0.42) = 33.1$$

so

$$h_f = \left(4\frac{f \Delta x_{\text{total}}}{D} + \sum_{\substack{\text{all valves} \\ \text{and fittings}}} K\right)\left(\frac{V^2}{2g}\right)$$

$$= \left(4\frac{0.004 \cdot 90\,\text{ft}}{\left(\dfrac{2}{12}\,\text{ft}\right)} + 33.1\right)\left(\frac{\left(10.2\,\text{ft/s}\right)^2}{2 \cdot 32.2\,\text{ft/s}^2}\right) = 67.4\,\text{ft}$$

Types of Fluid Flow Problems

The mathematical approach that must be conducted to get a solution may be different depending on the actual question or problem posed. The mathematical approach depends on what parameters are known and/or fixed. Three types of problems typically exist:

- Type 1: The flow rate and pipe diameter are known (or fixed), and the friction loss is unknown.
- Type 2: The pipe diameter and friction loss are known, and the flow rate is unknown.
- Type 3: The flow rate and friction loss are known, and the pipe diameter is unknown.

The common approaches to solve each type of fluid flow problem are described next.

Type 1 Approach

Known: flow rate and pipe diameter.
Unknown: friction loss.

Solution

1. Calculate fluid velocity V [$V = (4 \cdot Q)/(\pi \cdot D^2)$ for a pipe with a circular cross section].
2. Calculate Reynolds number $\Re$ with Eq. 5-4.
3. Calculate the relative roughness ε/D.
4. Find the friction factor f from the chart of pipe friction factors (Fig. 5-6) or one of the friction factor equations (e.g., the Colebrook equation, Eq. 5-48).
5. Find Ks for all valves and fittings from published tables or manufacturers' data.
6. Calculate the friction loss with Eqs. 5-72 or 5-73.

Type 2 Approach

Known: pipe diameter and friction loss.
Unknown: flow rate.

Solution

1. Assume (guess) a flow rate.
2. Calculate the friction loss as for type 1.

3. If the calculated friction loss is equal to the known friction loss, *stop*, as the answer has been achieved. If not, adjust the assumed (guessed) flow rate, and go back to step 2 to recalculate the friction loss at the new guessed flow rate. Repeat until the solution is reached.

Type 3 Approach

Known: flow rate and friction loss.
Unknown: pipe diameter.

Solution

1. Assume (guess) a pipe diameter.
2. Calculate the friction loss as for type 1.
3. If the calculated friction loss is equal to the known friction loss, *stop*, as the answer has been achieved. If not, adjust the assumed (guessed) pipe diameter, and go back to step 2 to recalculate the friction loss at the new guessed pipe diameter. Repeat until the solution is reached.

Example

The water storage tank in Fig. 5-17 discharges 16 °C water through a commercial steel pipe. The pipe is straight and 60 m in length, and has a square-edged inlet and a discharge. What pipe inner diameter is necessary to achieve 3.0 m³/min flow rate?

Solution

This is a type 3 problem with a known flow rate (3.0 m³/min) and friction loss (12 m), and the pipe diameter unknown. The problem must be solved by initially choosing a pipe diameter, and then calculating the head (friction) loss. If the head

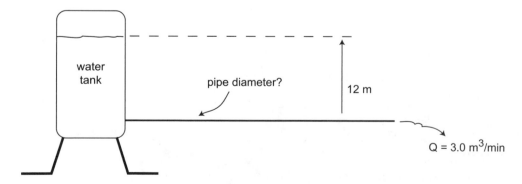

Figure 5-17. Schematic for type 3 flow example.

loss is greater than 12 m, the calculations must be done with a larger pipe diameter until the proper head loss is found. And vice versa.

Choose a pipe diameter of 1.0 m. The velocity in the pipe is then:

$$V = \frac{Q}{A} = \frac{3.0 \frac{m^3}{min}}{\frac{\partial}{4} \cdot (1.0 \text{ m})^2} \cdot \frac{1 \text{ min}}{60 \text{ s}} = 0.064 \frac{m}{s}$$

The sum of the friction loss coefficients (both values obtained from Table 5-2) is:

$$\Sigma K = K_{inlet} + K_{discharge} = 0.5 + 1.0 = 1.5$$

The Reynolds number is calculated to be:

$$\Re = \frac{D \cdot V \cdot \rho}{\mu} = \frac{1.0 \text{ m} \cdot 0.064 \frac{m}{s} \cdot 998.95 \frac{kg}{m^3}}{1.108 \times 10^{-3} \frac{kg}{m \cdot s}} = 5.77 \times 10^4$$

The friction factor, f, calculated from the Colebrook equation, is 5.11×10^{-3}. The head loss is:

$$h_f = \left(4 \frac{f \Delta x_{total}}{D} + \sum_{\substack{\text{all valves} \\ \text{and fittings}}} K \right) \left(\frac{V^2}{2g} \right)$$

$$h_f = \left(4 \frac{5.11 \times 10^{-3} \cdot 60 \text{ m}}{1.0 \text{ m}} + 1.5 \right) \left(\frac{\left(0.064 \frac{m}{s} \right)^2}{2 \cdot 9.81 \frac{m}{s^2}} \right) = 5.69 \times 10^{-4} \text{ m}$$

This friction loss is much less than the available friction loss, so the pipe size must be reduced. By trial and error, a pipe diameter of 0.116 m is found that satisfies the problem, specifically where the calculated head loss is equal to the available head loss.

Non-Newtonian Fluids

As discussed in Chapter 2, non-Newtonian fluids do not follow Newton's law of viscosity. Specific types of non-Newtonian fluids are Bingham plastics, pseudoplastics, and dilatants. Their behavior under shear is illustrated in Fig. 5-18.

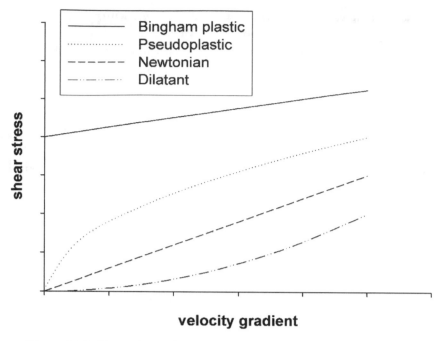

Figure 5-18. Newtonian and non-Newtonian fluid shear stress versus velocity gradient.

The power-law model is typically used to quantitatively describe the flow behavior of these non-Newtonian fluids:

$$\tau = m\left(\frac{dV}{dy}\right)^{n} \tag{5-74}$$

where m and n are properties of the fluid and dV/dy is the local velocity gradient. The power-law model describes the shear stress as a function of a constant m and the velocity gradient raised to a power n.

This can be rewritten as

$$\tau = m\left(\frac{dV}{dy}\right)^{n-1}\frac{dV}{dy} \tag{5-75}$$

where the term $m(dV/dy)^{n-1}$ is often called the "apparent viscosity," yet it is not independent of the velocity gradient.

The power law does not have a theoretical basis, but it does fit much of the experimental data; so it is empirically based. The value for n is less than one for pseudoplastics and greater than one for dilatants. Notice that the power law equation collapses to Newton's law of viscosity for $n = 1$ and $m = \mu$.

Because the behavior of non-Newtonian fluids under shear is different from that of Newtonian fluids, the velocity profile is different also. The equation describing the velocity profile for laminar flow of non-Newtonian fluids in a circular conduit of radius r_0 can be derived by a similar method to that used to find the velocity profile for laminar flow Newtonian fluids (derived earlier) and is as follows (McCabe et al. 1993):

$$V = \left(\frac{\tau_w g_c}{r_0 m} \right)^{1/n} \frac{r_0^{1+1/n} - r^{1+1/n}}{1+1/n} \tag{5-76}$$

where τ_w is the fluid shear at the pipe wall.

The relative velocities as a function of distance from the centerline of round conduit for dilatant, Newtonian, and pseudoplastic fluids using this equation are plotted in Fig. 5-19. The figure shows that the profile for a dilatant ($n > 1$) is more "pointed" than for Newtonian fluids, whereas the profile for a pseudoplastic ($n < 1$) is "flatter" or more "pluglike." The flow profiles for non-Newtonian fluids do not follow the parabolic profile that laminar Newtonian fluids do unless n is close to one.

Because the flow profiles for non-Newtonian fluids are different from those for Newtonian fluids, we may expect the friction losses to be different as well. In the same way that the Hagen–Poiseuille equation was derived for laminar flow through a balance on a fluid element, we can derive an equation that describes the pressure drop for laminar, non-Newtonian, power-law flow (McCabe et al. 1993):

$$\Delta P = \frac{2m}{g_c} \left(\frac{3n+1}{n} \right)^n \frac{V^n}{r_0^{n+1}} \Delta x \tag{5-77}$$

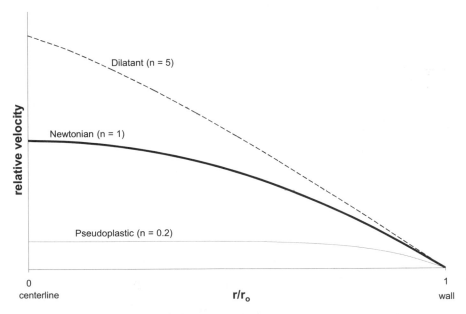

Figure 5-19. Laminar flow profile for Newtonian and non-Newtonian fluids (with identical τ_w and m) in round conduit.

Example

A non-Newtonian fluid with a 2.0×10^{-3} m³/s flow rate passes through a 5 cm diameter, 25 m long straight pipe. Laminar flow conditions exist, and m = 0.35 and m = 20 Pa·s$^{0.35}$. What is the pressure loss?

Solution

The fluid velocity is

$$\frac{Q}{\frac{\pi}{4}d^2} = \frac{2.0 \times 10^{-3} \frac{m^3}{s}}{\frac{\pi}{4}(.050 \text{ m})^2} = 1.02 \frac{m}{s}$$

Substituting values into Eq. 5-77 gives:

$$\Delta P = 2 \cdot 20 \text{ Pa} \cdot s^{0.35} \cdot \left[\frac{(3 \cdot 0.35) + 1}{0.35} \right]^{0.35} \cdot \frac{1.02 \frac{m}{s}}{(0.025 \text{ m})^{0.35+1}} \cdot 25 \text{ m} = 2.7 \times 10^5 \text{ Pa}$$

As for Newtonian flow, we can apply Bernoulli's equation to find

$$\Im = \frac{\Delta P}{\rho}$$

Substituting ΔP from Eq. 5-77 we get

$$\Im = \frac{\Delta P}{\rho} = \frac{\frac{2m}{g_c}\left(\frac{3n+1}{n}\right)^n \frac{V^n}{r_0^{n+1}} \Delta x}{\rho} \tag{5-78}$$

We can define the friction factor as before:

$$f = \frac{\Im}{4\left(\frac{\Delta x}{D}\right)\left(\frac{V^2}{2}\right)}$$

Substituting $\Im$ from Eq. 5-78 and simplifying yields for laminar flow

$$f = \frac{2^{n+1}m}{D^n \cdot \rho \cdot V^{2-n}}\left(3 + \frac{1}{n}\right)^n \tag{5-79}$$

Now apply the relation derived before for laminar Newtonian flow,

$$f = \frac{16}{\Re}$$

and combine with Eq. 5-79 to get

$$f = \frac{16}{\Re} = \frac{2^{n+1} m}{D^n \cdot \rho \cdot V^{2-n}} \left(3 + \frac{1}{n} \right)^n \qquad (5\text{-}80)$$

After simplifying, we have Reynolds number for non-Newtonian flow which can be applied to both laminar and turbulent flow:

$$\Re = 2^{3-n} \left(\frac{n}{3n+1} \right)^n \frac{D^n \tilde{n} V^{2-n}}{m} \qquad (5\text{-}81)$$

Example

Calculate the Reynolds number for the previous example problem where non-Newtonian fluid is passing through a 5 cm diameter pipe at a velocity of 1.02 m/s. Assume a fluid density of 1100 kg/m^3, and take $n = 0.35$ and $m = 20$ Pa·s$^{0.35}$.

Solution

Substituting values into Eq. 5-81:

$$\Re = 2^{(3-0.35)} \cdot \left[\frac{0.35}{(3 \cdot 0.35) + 1} \right]^{0.35} \cdot \frac{(0.05 \text{ m})^{0.35} \cdot 1100 \frac{\text{kg}}{\text{m}^3} \cdot \left(1.02 \frac{\text{m}}{\text{s}} \right)^{(2-0.35)}}{20 \frac{\text{kg} \cdot \frac{\text{m}}{\text{s}^2}}{\text{m}^2} \cdot \text{s}^{0.35}} = 67$$

For $n = 1$ and $m = \mu$, this equation for Reynolds number reduces to that for a Newtonian fluid. Using the non-Newtonian Reynolds number calculated with Eq. 5-81, the friction factor can be determined for laminar flow by using Eq. 5-82:

$$f = \frac{16}{\Re} \qquad (5\text{-}82)$$

For non-Newtonian turbulent flow, the following correlation has been developed (Dodge and Metzner 1959):

$$\frac{1}{\sqrt{f}} = \frac{4.0}{n^{0.75}} \cdot \log \left[\Re \cdot f^{\left(1 - \frac{n}{2}\right)} \right] - \frac{0.4}{n^{1.2}} \qquad (5\text{-}83)$$

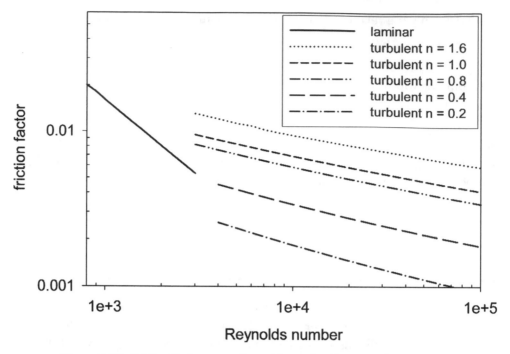

Figure 5-20. Friction factor versus Reynolds number for power-law non-Newtonian fluids. Friction factor calculated from Eqs. 5-82 and 5-83. *Source:* Adapted from Dodge and Metzner (1959) and McCabe et al. (1993).

Equations 5-82 and 5-83 are plotted in Fig. 5-19. From the friction factor, the friction loss of the non-Newtonian fluid, flowing either laminar or turbulent, can be calculated from

$$\mathfrak{I} = \frac{\Delta P}{\rho} = 4f\left(\frac{\Delta x}{D}\right)\left(\frac{V^2}{2}\right)$$

as presented earlier in the chapter.

Symbol List

C_V	flow coefficient
d	internal diameter
D	pipe diameter
D_{eq}	equivalent diameter
f	friction factor
$\mathfrak{I}$	frictional loss
g_c	Newton's law proportionality factor, 32.174 ft·lbm/lbf·s^2
h_f	frictional head loss

K	fitting resistance coefficient
L	characteristic length
L_p	wetted perimeter
m, n	non-Newtonian fluid power-law coefficients
P	pressure
r	radius
r_0	pipe radius
r_H	hydraulic radius
$\Re$	Reynolds number
S	cross-sectional area
SG	specific gravity
V	fluid velocity, average fluid velocity
W	work
Δx	conduit length
x_t	transition length
ε	surface roughness
ρ	fluid density
τ	shear stress
τ_w	fluid shear at pipe wall
μ	absolute fluid viscosity

Problems

1. A 1.0-in. inner diameter cast iron pipe carries 0.5 gpm of 40 °F (5 °C) water for 125 ft. Is the flow laminar or turbulent? The fittings in the pipe include a square-edged inlet, a discharge, and two regular screwed 90° elbows. What is the difference in pressure between the inlet and exit of the pipe if they are at the same static height?

2. What is the flow rate of 20 °C water through a 300-mm-diameter, clean (new) cast iron pipe that is 500 m long with a 0.4-m head difference between the inlet and outlet to the pipe? The pipe is straight with no valves.

3. The surface roughnesses listed in Table 5-1 are for new, clean pipe. The roughness can be expected to increase over time because of corrosion and scaling (see Chapter 10). What would the expected flow rate be for the pipe in Problem 2 with a surface roughness that is 20 times the value for new, clean pipe? What pipe inner diameter should be used to maintain the same flow rate achieved in Problem 2 with the same head loss? Adapted from Problem 3.6 of Wilkes (1999).

4. A 2.0-in.-diameter, 100-ft-long pipe carries a non-Newtonian fluid at 200 gpm. The fluid density is 62.2 lb/ft^3 and power-law parameters n and m are 0.3 and 5.5 Pa·s$^{0.3}$, respectively. What is the friction loss in pounds per square inch in the pipe?

5. You are transferring the liquids from the tank in Problem 3 of Chapter 3 through a piping system that is connected at the bottom of the tank. The tank

always remains full of water and/or gasoline. When flow is initiated, water flows though the transfer piping. After the water has been removed (and the tank is full of gasoline), gasoline flows through the transfer piping. The piping is commercial steel, with a 60-mm inner diameter and 50 m in length. Assume that there are no fittings or valves and that the frictional loss is only due to pipe friction and remains constant regardless of the liquid flowing through the piping. If a flow rate of 4 m³/hr is achieved when water flows through the piping, what flow rate should be expected when gasoline flows? Assume the viscosity of gasoline is 0.5 cP. What would the expected flow rate through the transfer piping be if fuel oil was being separated from water in the tank and then transferred (in lieu of gasoline)? Assume that the fuel oil has a viscosity of 3.0 cP and a specific gravity of 0.87.

References

Avallone, E. A., and Baumeister, T., III, eds. (1996). *Marks' Standard Handbook for Mechanical Engineers*, McGraw-Hill, New York.

Crane Co. (1985). "Flow of Fluids through Valves, Fittings, and Pipe," *Tech. Paper No. 410*, Crane Co., New York.

Dodge, D. W., and Metzner, A. B. (1959). "Turbulent Flow of Non-Newtonian Systems," *Am. Inst. Chem. Eng. J.* 5, 189–204.

Hydraulic Institute (1961). *Pipe Friction Manual*, Hydraulic Institute, Parsippany, NJ.

Hydraulic Institute (1990). *Engineering Data Book*, Hydraulic Institute, Parsippany, NJ.

Levi, E. (1995). *The Science of Water. The Foundation of Modern Hydraulics*, ASCE Press, New York.

McCabe, W. L., Smith, J. C., and Harriott, P. (1993). *Unit Operations of Chemical Engineering*, McGraw-Hill, New York.

Wilkes, J. O. (1999). *Fluid Mechanics for Chemical Engineers*, Prentice Hall, Upper Saddle River, NJ.

CHAPTER 6

Pumps and Motors

Chapter Objectives

1. Identify different pump types used in treatment systems.
2. Discuss centrifugal pump theory.
3. Recall centrifugal pump classifications.
4. Interpret pump characteristic curves and find system operating points.
5. Define cavitation and net positive suction head.
6. Recall the different electric motors used for driving pumps.

Typically in treatment systems, pumps are used to transfer fluids from one location to another. Pumps are used to move water, wastewater, and sludge as well as nonaqueous fluids such as gasoline, fuel oil, acids, and bases. Small metering pumps such as shown in Fig. 6-1 are employed, as well as larger centrifugal pumps, as illustrated in Fig. 6-2.

Pumps impart energy to fluids that can be converted to other energy forms. The energy transferred to the fluid can produce an increased pressure (P), height (z), or velocity (V) or overcome friction loss ($\Im$). Writing Bernoulli's equation across a pump, we start with

$$\Delta\left(\frac{P}{\rho} + gz + \frac{V^2}{2}\right) = \frac{-dW}{dm} - \Im \tag{6-1}$$

where dW is the work done by the fluid, or $-dW$ is the work done to the fluid by a pump.

The head form of Bernoulli's equation can be written as

$$\Delta\left(\frac{P}{\rho g} + z + \frac{V^2}{2g}\right) = \frac{-dW}{g(dm)} - \frac{\Im}{g} \tag{6-2}$$

Rearranging the head form of Bernoulli's equation gives

$$\frac{-dW}{g(dm)} = \frac{\Im}{g} + \Delta\left(\frac{P}{\rho g} + z + \frac{V^2}{2g}\right) \tag{6-3}$$

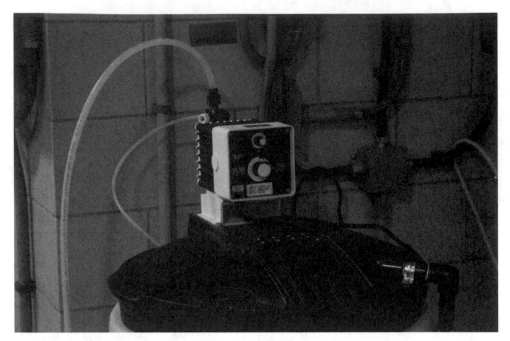

Figure 6-1. Chemical metering pump supplying chemicals at a water treatment plant.

Figure 6-2. Centrifugal pump feeding an ultrafiltration system in a drinking water treatment facility.

This may be simplified to

$$\frac{-dW}{g(dm)} = H_t \tag{6-4}$$

where

$$H_t = \frac{\Im}{g} + \Delta\left(\frac{P}{\rho g} + z + \frac{V^2}{2g}\right) \tag{6-5}$$

H_t is the *total dynamic head* that a pump must work against; it includes the friction head loss ($\Im/g$) and head from the pressure difference ($\Delta P/\rho g$), static height difference (Δz), and the difference in velocity heads ($\Delta V^2/2g$). Note that dW is negative, as work is being done *to* the fluid.

We can rearrange $[-dW/g(dm)] = H_t$ to

$$-dW = g \cdot dm \cdot H_t \tag{6-6}$$

and divide by dt to yield

$$-\frac{dW}{dt} = g \cdot \frac{dm}{dt} \cdot H_t \tag{6-7}$$

The power expended by a pump is the rate at which work is done:

$$P = \frac{dW}{dt} \tag{6-8}$$

Combining Eq. 6-7 and Eq. 6-8 gives P, the power delivered to the liquid:

$$P = g \cdot \dot{m} \cdot H_t \tag{6-9}$$

where $\dot{m}$ is the mass flow rate of the fluid with units of mass per time, which is equal to dm/dt for constant mass flow rate. Note that the minus sign is removed per convention, since the direction of energy transfer is to the fluid and is obvious at this point.

In terms of volumetric flow rate of the fluid the power is

$$P = g \cdot \rho \cdot Q \cdot H_t \tag{6-10}$$

Common units for power are kilowatts (kW or 1000 J/s) and horsepower (hp). Conversions among power units are as follows:

$$1 \text{ hp (horsepower)} = 550 \text{ ft} \times \text{lbf/s}$$
$$1 \text{ hp} = 2545 \text{ Btu/hr}$$
$$1 \text{ hp} = 0.7457 \text{ kW}$$

Not all power applied to the pump input shaft gets transferred to the fluid. A portion is lost in heat through friction in bearings, through the pump casing, via excessive turbulence, etc. The efficiency of power delivery, or the pump efficiency, is expressed as

$$E_P = \frac{P}{P_{shaft}} \tag{6-11}$$

where P is the power delivered to the liquid, which is sometimes called the hydraulic power (or hydraulic horsepower, water horsepower), and P_{shaft} is the power delivered to the shaft (sometimes called brake horsepower).

Equation 6-11 can be rearranged to solve for P_{shaft}, the power that needs to be delivered to the pump:

$$P_{shaft} = \frac{P}{E_P} \tag{6-12}$$

Substituting for P (defined by Eq. 6-10) we have

$$P_{shaft} = \frac{g \cdot \rho \cdot Q \cdot H_t}{E_P} \tag{6-13}$$

Equation 6-13 can be simplified to a dimensional equation (one with required units for each variable) for the U.S. Customary System:

$$P_{shaft} = \frac{\text{gpm} \cdot H_t \cdot SG}{3960 \cdot E_P} \tag{6-14}$$

Example

A pump is used to transfer 14 m³/min of 12 °C water to a tank at a greater elevation, as shown in Fig. 6-3. The frictional loss in the suction line is 0.10 m and in the discharge line it is 1.0 m. What are the total dynamic head and the power delivered to the water? If the efficiency of the pump is 60%, what power must be delivered to the pump shaft?

We start with Bernoulli's equation

$$H_t = \frac{\Im}{g} + \Delta \left(\frac{P}{\rho g} + z + \frac{V^2}{2g} \right)$$

$$= (0.10\,\text{m} + 1.0\,\text{m}) + \left[\left(\frac{P}{\rho g} + z + \frac{V^2}{2g} \right)_{discharge} - \left(\frac{P}{\rho g} + z + \frac{V^2}{2g} \right)_{suction} \right]$$

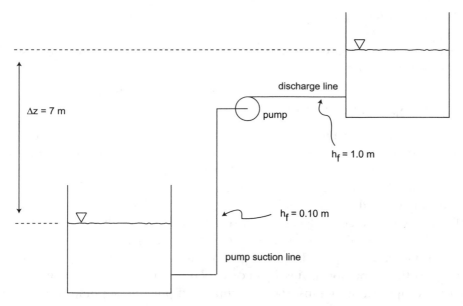

Figure 6-3. Schematic of piping configuration for pump example.

The pressures P on the discharge and suction tank surfaces are atmospheric and cancel. The velocities V at the tank surfaces are negligible and may be dropped from the equation. So the total dynamic head equation simplifies to

$$H_t = (0.10 \text{ m} + 1.0 \text{ m}) + [(z)_{\text{discharge}} - (z)_{\text{suction}}]$$

$$= 1.1 \text{ m} + 7 \text{ m} = 8.1 \text{ m}$$

The power delivered to the water (P or whp) is

$$P = g \cdot \rho \cdot Q \cdot H_t$$

and so

$$P = 9.81 \, ^m\!/_{s^2} \cdot 999.50 \, ^{kg}\!/_{m^3} \cdot 14 \, ^{m^3}\!/_{min} \cdot 8.1 \text{ m} \cdot \frac{1 \text{ min}}{60 \text{ s}}$$

$$= 1.85 \times 10^4 \, ^{N \cdot m}\!/_{s} = 18.5 \text{ kW}$$

To calculate the power that must be delivered to the pump, use Eq. 6-12, which gives

$$P_{\text{shaft}} = \frac{P}{E_P} = \frac{18.5 \text{ kW}}{0.60} = 30.8 \text{ kW} \approx 31 \text{ kW}$$

General Pump Types

The two general pump classifications—positive displacement pumps and kinetic pumps—are shown in Fig. 6-4.

Positive Displacement Pumps

Positive displacement pumps have a component that moves to "positively" displace fluid. In the displacement of fluid, the pressure is increased as the fluid passes through the pump. Positive displacement pumps may be relatively low capacity (low flow rate) and produce a high discharge pressure. They can be used for moving many different liquids, but they are particularly suited for liquids with solids in them (sewage, sludge, slurries, and the like), although kinetic pumps may be used with these fluids (sometimes successfully and sometimes not). The flow rate through a positive displacement pump is affected very little by the pressure (or head) that it pumps against—hence explaining the term "positive displacement." Reciprocating pumps, rotary pumps, and pneumatic pumps are all positive displacement pumps.

A plunger pump is a type of reciprocating pump. Plunger pumps are commonly used for clarifier underflow, sewage, sludge, and similar liquids containing solids. Plunger pumps can have one or multiple plungers in mating cylinders, driven by a rotating crankshaft. The delivered flow rate depends on the stroke, plunger diameter (with the stroke and plunger diameter defining the swept volume), and the rotational velocity of the crankshaft. The flow through the pump is maintained in one direction through the use of check valves, commonly ball check valves. Often dual check valves are used on both suction and discharge (for a total of four valves) to prevent pump malfunction in the event of one valve being held open with debris. The second valve will seal the reverse flow and the fluid flow will eventually clear the clogged valve.

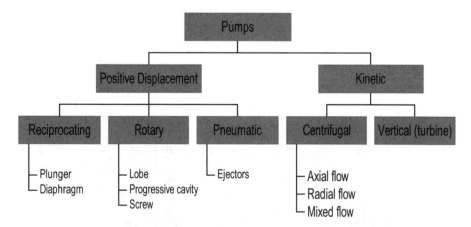

Figure 6-4. Pump classifications. *Source*: Adapted from Hydraulic Institute (1975).

A lobe (rotary) pump is shown in Fig. 6-5. The rotors are commonly coated with an elastomer such as urethane or nitrile. Clearance between the rotors is small enough so that liquid does not escape between the rotors and large enough to prevent rotor-to-rotor contact. The liquid thus moves in the direction that the outer lobe turns and is directed to the discharge of the pump. The rotors can have bearings on both ends of the shaft, or they can be in a cantilever arrangement with the bearings located on one end and the rotor on the other end. The discharge from a lobe pump is continuous, as opposed to a plunger pump where the discharge alternates between each plunger (cylinder). Lobe pumps are self-priming, and relatively unaffected by solid materials such as rags and larger solids because of smooth flow paths and no check valves. If a large piece of debris does get caught, the direction of rotation can be reversed to dislodge the debris. These pumps can be used for transferring sludge with up to 6% solids.

A schematic of a pneumatic ejector is shown in Fig. 6-6. Pneumatic ejectors transfer liquids through two cycles:

1. the filling cycle, during which the liquid flows into a tank, and
2. the ejection cycle, during which the tank is subjected to high pressure with compressed air.

Pneumatic ejectors are appropriate for low flow rates, up to approximately 2.5 m³/ min (660 gpm), and they can operate at discharge heads up to 100 m (330 ft). Because of exposure to air, freezing can be a problem in cold climates.

Centrifugal Pumps

Centrifugal pumps are pumps in which energy is transferred to the fluid by an impeller rotating at a fixed rotational velocity. The kinetic energy in the rotating impeller or impellers can move the fluid and impart increased pressure. Centrifugal pumps may have one impeller, in which case they are termed single-stage pumps,

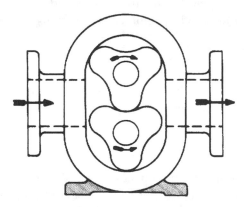

Figure 6-5. Drawing of a lobe (rotary) pump. *Source:* Courtesy of Hydraulic Institute, Parsippany, NJ 07054; http://www.pumps.org.

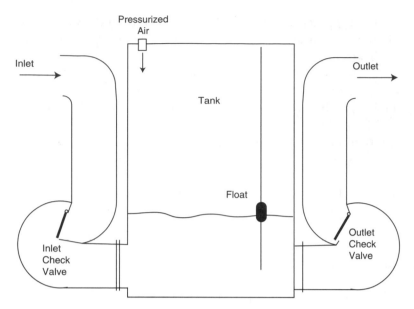

Figure 6-6. Schematic of a pneumatic ejector.

or they may have more than one impeller in series in the same pump casing (multistage pumps). Horizontal centrifugal pumps are oriented with the shaft horizontal as in Fig. 6-2, and vertical centrifugal pumps have vertical shafts as in Fig. 6-7. Centrifugal pumps are classified according to the type of flow (radial, axial, or mixed) in the pumps, which is characteristic of the impeller and casing design.

Radial Flow Centrifugal Pumps

In a radial flow centrifugal pump, energy is transferred from the impeller to the fluid through centrifugal force from the impeller in the radial direction, as shown in Fig. 6-8. The liquid enters the impeller through the suction flange, the hub (inner portion of the impeller), and then flows radially outward by centrifugal force. Single suction and double suction configurations are used.

Axial Flow Centrifugal Pumps

In an axial flow centrifugal pump, energy is transferred to the fluid from the lifting action of vanes in an axial flow pump. The vanes function as "propellers." The flow is predominantly in the same direction as the axis of rotation of the impeller, as shown in Fig. 6-9.

Mixed Flow Centrifugal Pumps

In a mixed flow centrifugal pump, energy is transferred to the fluid by a combination of centrifugal force and lift. The flow enters the single suction impeller axially and discharges at an angle to the rotational axis, as shown in Fig. 6-10. The impeller discharges part way between complete axial and radial directions.

Figure 6-7. Vertical centrifugal pumps.

Centrifugal Pump Configurations

As discussed earlier, centrifugal pumps are of three types: radial, axial, and mixed flow. Centrifugal pumps may be configured with single suction impellers, with one inlet eye on one side of the impeller hub (see Fig. 6-8), or double suction impellers with two inlet eyes, one on each side of the impeller (see Fig. 6-11). Figure 6-12 illustrates a multistage radial flow centrifugal pump where the discharge from the first stage enters the impeller suction eye of the second stage, etc. A mixed flow vertically oriented (vertical pump shaft) centrifugal pump is depicted in Fig. 6-13.

Rolling element bearings are usually used to support the shaft on which the impeller is mounted. The bearings carry radial and thrust loads, transferring the loads from the shaft to the housing. The design life of rolling element bearings is limited by the eventually failure due to surface fatigue (flaking, spalling) of the surfaces that repeatedly contact as the shaft rotates. Specifically, it is expected that surface fatigue will eventually cause failure of the surfaces of the rolling elements and the races that the elements contact. However, it has been found that a significant amount (up to 75%) of installed bearings prematurely fail due to contamination in the bearing before the expected design life is achieved. This contamination in the bearing internals comes about from seal failure and produces rust, scratching, and pitting from foreign material between the rolling elements and the races. This damage to the bearing surfaces is usually quickly followed by bearing failure.

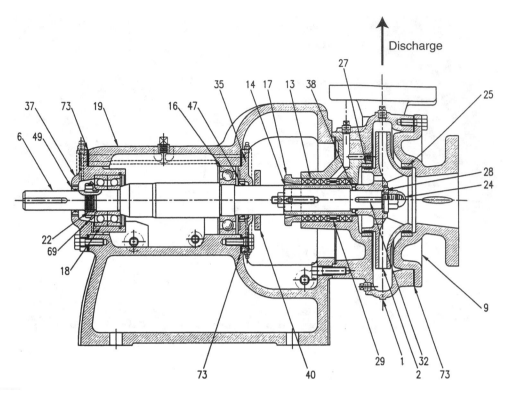

Figure 6-8. Schematic of a radial flow pump.

Key: 1—casing, 2—impeller, 6—shaft, 9—cover, 13—packing, 14—sleeve, 16—bearing, 17—gland, 18—bearing, 19—frame, 22—bearing locknut, 24—impeller nut, 25—cover ring, 27—stuffing-box cover ring, 28—gasket, 29—lantern ring, 32—impeller key, 35—bearing cover, 37—bearing cover, 38—shaft sleeve gasket, 40—deflector, 47—bearing cover seal, 49—bearing cover seal, 69—lock washer, 73—gasket, 78—bearing spacer.

Source: Courtesy of Hydraulic Institute, Parsippany, NJ 07054; http://www.pumps.org.

Centrifugal Pump Principles

As fluid gains motion by contact with an impeller rotating at an angular velocity ω, the fluid entering at the hub with a radius r_1 at velocity V_1 is accelerated to V_2 at radius r_2 (where subscript 1 refers to the inner hub radius and subscript 2 refers to the outer diameter of the impeller). See Fig. 6-14.

A balance of angular momentum produces the equation

$$T = \frac{\gamma}{g} Q \left[r_2 V_2 \cos \alpha_2 - r_1 V_1 \cos \alpha_1 \right] \tag{6-15}$$

where T is torque, γ is the fluid specific weight, Q is the volumetric fluid flow rate, α_1 is the angle of V_1 to the tangent of the inner radius of the impeller, r_1, and α_2

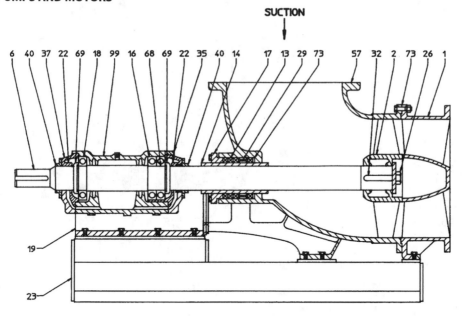

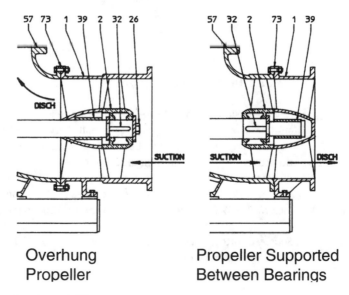

Overhung
Propeller

Propeller Supported
Between Bearings

Figure 6-9. Schematic of an axial flow pump.

Key: 1—casing, 2—impeller, 6—shaft, 13—packing, 14—shaft sleeve, 16—bearing, 17—gland, 18—bearing, 19—frame, 22—bearing locknut, 23—base plate, 26—impeller screw, 29—lantern ring, 32—impeller key, 35—bearing cover, 37—bearing cover, 39—bushing, 40—deflector, 57—suction elbow, 68—shaft collar, 69—lock washer, 73—gasket, 99—bearing housing.

Source: Courtesy of Hydraulic Institute, Parsippany, NJ 07054; http://www.pumps.org.

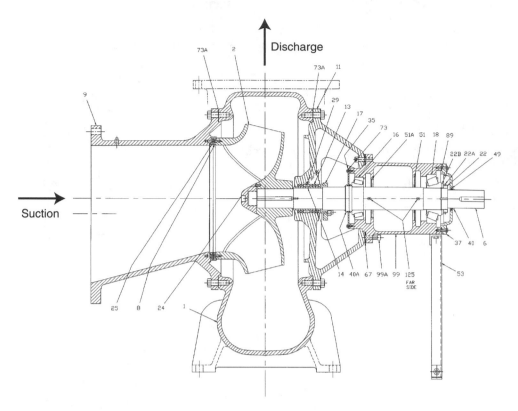

Figure 6-10. Schematic of a mixed flow pump.

Key: 1—casing, 2—impeller, 6—shaft, 8—impeller ring, 9—cover, 11—stuffing-box cover, 13—packing, 14—shaft sleeve, 16—bearing, 17—gland, 18—bearing, 19—frame, 22—bearing locknut, 24—impeller nut, 25—suction cover ring, 29—lantern ring, 32—impeller key, 33—bearing housing, 35—bearing cover, 38—shaft sleeve gasket, 40—deflector, 43—bearing cap, 47—bearing cover seal, 49—bearing cover seal, 67—frame liner shim, 69—lock washer, 73—gasket, 78—bearing spacer, 169—bearing housing seal.

Source: Courtesy of Hydraulic Institute, Parsippany, NJ 07054; http://www.pumps.org.

is the angle of V_2 to the tangent of the outer radius of the impeller, r_2. Multiplying the torque on the impeller by the angular velocity gives the power:

$$P = T \cdot \omega = \gamma \cdot Q \cdot H \tag{6-16}$$

where H is the head added to the fluid and ω is the rotational velocity. So the torque is

$$T = \frac{\gamma \cdot Q \cdot H}{\omega} \tag{6-17}$$

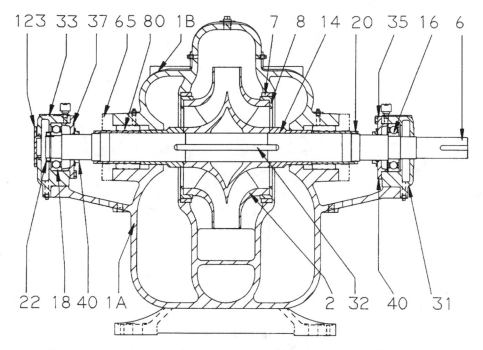

Figure 6-11. Schematic of a double suction radial flow pump.

Key: 1A—lower casing, 1B—upper casing, 2—impeller, 6—shaft, 7—casing ring, 8—impeller ring, 14—shaft sleeve, 16—bearing, 18—bearing, 20—shaft sleeve nut, 22—locknut, 31—bearing housing, 32—impeller key, 33—bearing housing, 35—bearing cover, 37—bearing cover, 40—deflector, 65—mechanical seal, stationary element, 80—mechanical seal, rotating element, 123—bearing end cover.

Source: Courtesy of Hydraulic Institute, Parsippany, NJ 07054; http://www.pumps.org.

Substituting Eq. 6-17 for T into Eq. 6-15 yields

$$\frac{\gamma}{g} Q \left[r_2 V_2 \cos \alpha_2 - r_1 V_1 \cos \alpha_1 \right] = \frac{\gamma \cdot Q \cdot H}{\omega} \qquad (6\text{-}18)$$

Solving for H, we get

$$H = \frac{\left[\omega \cdot r_2 V_2 \cos \alpha_2 - \omega \cdot r_1 V_1 \cos \alpha_1 \right]}{g} \qquad (6\text{-}19)$$

For radial flow pumps, $\cos \alpha_1$ can be taken to be 0 and so

$$H = \frac{\omega \cdot r_2 V_2 \cos \alpha_2}{g} \qquad (6\text{-}20)$$

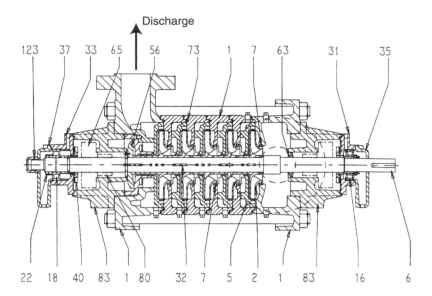

Figure 6-12. Schematic of a multistage radial flow centrifugal pump.

Key: 1—casing, 2—impeller, 5—diffuser, 6—shaft, 7—casing ring, 16—bearing, 18—bearing, 22—bearing locknut, 31—bearing housing, 32—impeller key, 33—bearing housing, 35—bearing cover, 37—bearing cover, 40—deflector, 56—balancing disk, 63—stuffing-box bushing, 65—mechanical seal, stationary element, 73—gasket, 80—mechanical seal, rotating element, 83—stuffing-box, 123—bearing end cover.

Source: Courtesy of Hydraulic Institute, Parsippany, NJ 07054; http://www.pumps.org.

Therefore the head that a centrifugal pump can theoretically produce is independent of the fluid density (specific weight) and viscosity of the fluid. It is dependent on the fluid velocity at the outer diameter of the impeller, which is a function of both the rotational velocity and radius of the impeller. The impeller rotational velocity and impeller radius are important parameters that greatly affect the characteristics of centrifugal pumps.

Using dimensional analysis techniques, the following dimensionless groups can be developed for pumps:

$$C_Q = \frac{Q}{nD^3} \quad \text{(capacity coefficient)} \tag{6-21}$$

$$C_H = \frac{gH}{n^2 D^2} \quad \text{(head coefficient)} \tag{6-22}$$

$$C_P = \frac{P}{\rho n^3 D^5} \quad \text{(power coefficient)} \tag{6-23}$$

where n is the rotational velocity.

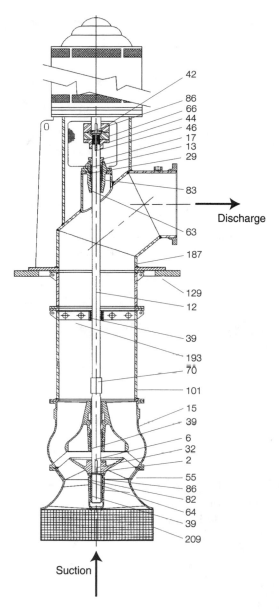

Figure 6-13. Schematic of a mixed flow vertical, wet-well, centrifugal pump.

Key: 2—impeller, 6—pump shaft, 12—line shaft, 13—packing, 15—discharge bowl, 17—gland, 29—lantern ring, 32—impeller key, 39—bushing, bearing, 42—driver coupling half, 44—pump coupling half, 46—coupling key, 55—suction bell, 63—stuffing-box bushing, 64—protecting collar, 66—shaft adjusting nut, 70—shaft coupling, 82—thrust ring retainer, 83—stuffing-box, 86—thrust split-ring, 101—column pipe, 129—sole plate, 187—surface discharge head, 193—bearing retainer, 209—strainer.

Source: Courtesy of Hydraulic Institute, Parsippany, NJ 07054; http://www.pumps.org.

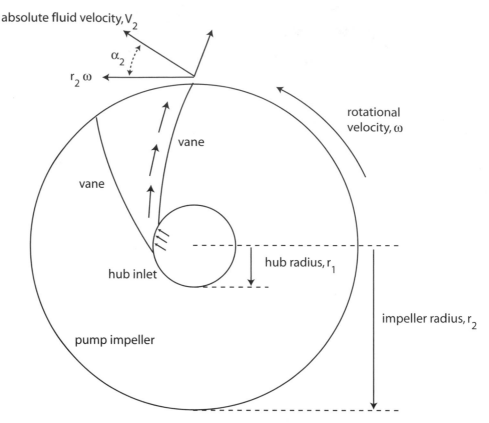

Figure 6-14. Fluid velocities in a radial flow pump. *Source:* Adapted from Sanks et al. (1998).

These dimensionless groups are themselves used very little, if at all. But they are derivable from dimensional analysis. Moreover, they have been combined into very useful relationships called *affinity laws*. These affinity laws may be used to estimate the effect of changing certain dimensions and parameters on the performance of a pump.

Pump Affinity Laws

By writing the capacity coefficient C_Q for two geometrically similar pumps with different flow rates Q_1 and Q_2, rotational velocities n_1 and n_2, and the same impeller diameter D, we obtain

$$C_Q = \frac{Q_1}{n_1 D^3} = \frac{Q_2}{n_2 D^3} \tag{6-24}$$

Rearranging and canceling D yields

$$\frac{Q_1}{Q_2} = \frac{n_1}{n_2} \tag{6-25}$$

Similarly, the following can be derived from writing C_H and C_P for two geometrically similar pumps:

$$\frac{H_1}{H_2} = \left(\frac{n_1}{n_2}\right)^2 \qquad (6\text{-}26)$$

$$\frac{P_1}{P_2} = \left(\frac{n_1}{n_2}\right)^3 \qquad (6\text{-}27)$$

These three equations are known as pump affinity laws. These three affinity laws show the effect of rotational velocity on flow rate, pump head, and pump power.

To provide the most flexible pump design, manufacturers may provide a range of impeller diameters for each casing. We know from Eq. 6-20, that impeller diameter is an important parameter for determining pump performance. We can use affinity laws to estimate the effect of changing impeller diameter. The capacity coefficient C_Q for two geometrically similar pumps with different impeller diameters and flow rates, and the same rotational velocity is

$$C_Q = \frac{Q_1}{nD_1^3} = \frac{Q_2}{nD_2^3} \qquad (6\text{-}28)$$

Simplifying and canceling n yields

$$\frac{Q_1}{Q_2} = \left(\frac{D_1}{D_2}\right)^3 \qquad (6\text{-}29)$$

However, from a dimensional analysis perspective this equation can be considered to be for a *three-dimensional scale-up* of a pump. The three dimensions of concern here are impeller diameter, impeller width, and volute size. If we hold impeller width and volute size constant, we obtain an equation for the *one-dimensional scale-up* based on the impeller diameter:

$$\frac{Q_1}{Q_2} = \frac{D_1}{D_2} \qquad (6\text{-}30)$$

Starting with C_H, we can similarly find

$$\frac{H_1}{H_2} = \left(\frac{D_1}{D_2}\right)^2 \qquad (6\text{-}31)$$

Because power is determined by the head and flow rate ($g \cdot \rho \cdot Q \cdot H$) we have

$$\frac{P_1}{P_2} = \frac{g\rho \cdot Q_1 H_1}{g\rho \cdot Q_2 H_2} = \frac{D_1}{D_2}\left(\frac{D_1}{D_2}\right)^2 \qquad (6\text{-}32)$$

or

$$\frac{P_1}{P_2} = \left(\frac{D_1}{D_2}\right)^3 \qquad (6\text{-}33)$$

These equations should be taken as approximations only, as scale-up of pumps is much more complicated than represented by these simple scaling formulas.

Example (from Karassik, 1981)

A 3500 rpm motor rated at 5 hp is coupled to a centrifugal pump with a 7 in. diameter impeller. It is to be installed in a system where it must produce 20 gpm at 90 psi discharge pressure. Before installation, the pump/motor is tested and it is found that the pump yielded 20 gpm at 100 psi with the motor developing 6 hp. The discharge pressure and the load on the motor must be reduced. What should the impeller diameter be changed to?

Solution

From Eq. 6-31:

$$\frac{H_1}{H_2} = \left(\frac{D_1}{D_2}\right)^2$$

Substituting in values:

$$\frac{900 \text{ psi}}{100 \text{ psi}} = \left(\frac{D_1}{7.0 \text{ in}}\right)^2$$

Rearranging to solve for D_1:

$$D_1 = 7.0 \text{ in} \left(\frac{90 \text{ psi}}{100 \text{ psi}}\right)^{1/2} = 6.64 \text{ in} \approx 6 \tfrac{5}{8} \text{ in}$$

The impeller diameter must be reduced to $6\,\tfrac{5}{8}$ in.

The developed power will be affected by the reduction in impeller diameter. Using Eq. 6-33:

$$\frac{P_1}{P_2} = \left(\frac{D_1}{D_2}\right)^3$$

Substituting in values:

$$\frac{P_1}{6 \text{ hp}} = \left(\frac{6.64 \text{ in}}{7.0 \text{ in}}\right)^3$$

Rearranging to solve for P_1 gives:

$$P_1 = 6 \text{ hp} \cdot \left(\frac{6.64 \text{ in}}{7.0 \text{ in}} \right)^3 = 5.1 \text{ hp}$$

So the power developed by the motor would be about equal to the rated power.

___

Pump Characteristic Curves

As flow rate though a centrifugal pump is changed, the total head produced by the pump may change also. The theoretical relationship between Q and H is elusive. Because of turbulence (and other energy losses) in the pump, the actual transfer of energy from kinetic to pressure (or head) energy is complicated and cannot be predicted very well. The head produced as a function of flow rate is usually measured. These measurements are supplied to the fluid system designer by the manufacturer in the form of *pump characteristic curves*.

Attempts have been made to quantify pump head versus flow rate with a mathematical equation. One commonly used empirical relationship is

$$H = a - bQ^c \tag{6-34}$$

where a, b, and c are empirical constants. The value for c is sometimes taken to be 2.

Pump characteristic curves are found through flow testing by the pump designer or manufacturer and are available to the fluid system designer. A manufacturer's pump curve is shown in Fig. 6-15.

For most centrifugal pumps, a pump curve does not necessarily "fix" the total dynamic head delivered by the pump nor the flow rate of the fluid through the pump. The piping system in which the pump is installed affects the pump operating point. For proper system operation, the total dynamic head provided by the pump must be equal to or greater than the head needed by the system. The head needed by the system is that due to static height differences and friction losses and is a function of flow rate through the system. The system curve can be created a priori by calculating the total system head at increasing flow rates to form the continuous curve. During actual system operation, the total dynamic head will be *equal* to the system head, as the system will find equilibrium at the system operating point, as shown in Fig. 6-16. The system head may be adjusted with throttling valves, or other methods of head (friction loss) control, to move the operating point of the system on the pump curve.

Multiple pumps are often installed in systems in series or in parallel. The total dynamic head produced by pumps in series is the sum of the total dynamic heads of the individual pumps at every flow rate on the pump curve. For identical pumps in parallel, the flow rates of the pumps are summed at every total dynamic head. See Fig. 6-17. Connecting pumps together in a system provides additional frictional losses from the connecting headers, which would not be present if only a

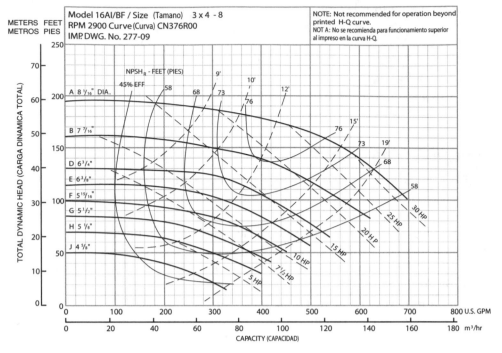

Figure 6-15. A centrifugal pump curve. *Source:* Courtesy Goulds Pumps, ITT Industries, Seneca Falls, NY.

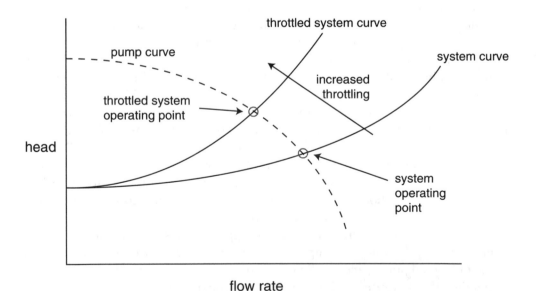

Figure 6-16. System operating point at intersection of pump curve and system curve.

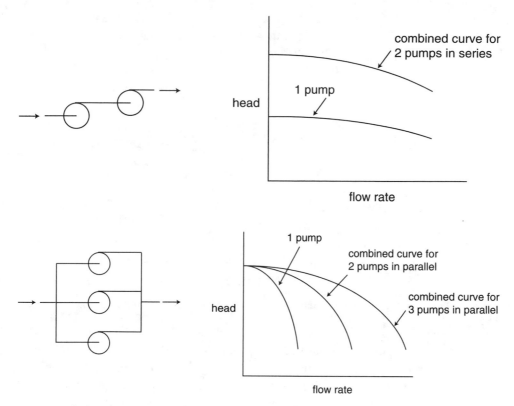

Figure 6-17. Pumps in series and parallel.

single pump were present. The presence of the connecting headers may produce significant friction losses and if so, the friction losses in the headers must be accounted for in system calculations. In some cases, due to the location of the pumps, the friction loss in the connecting headers may be different for each pump. A tactic to address this is to subtract the header friction loss from each pump to get an "effective" pump curve for each pump, and then sum the individual curves to get a combined "effective" pump curve that accounts for the different friction loss in each pump header. The intersection of this combined pump curve with the system curve will define the system operating point.

The system operating point for multiple pumps is located at the intersection of the system curve and the combined pump curve. Figure 6-18 shows the operating point for two nonidentical pumps in parallel. The two pump curves are summed to produce the combined pump curve, and the intersection with the system curve indicates the system operating point.

System Control

Often systems in which pumps operate have varying conditions that may affect the pumps. In a supply-controlled system, pumps may be transferring fluid from a

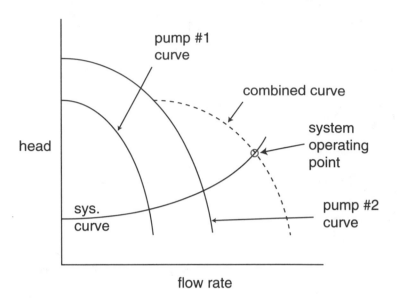

Figure 6-18. System operating point for two nonidentical pumps in parallel.

supply tank, and the tank level fluctuates as the pumps operate. In a demand-controlled system, pumps may transfer to a receiving tank, and the level of that tank may control the pumps (e.g., a drinking water storage tank). So the overall fluid static height that is pumped against may vary. In addition, the total friction loss in a system may change because of flow rate changes. This can be brought about via throttling valve adjustments. The engineer must determine the anticipated changes to the system curve and make sure that the pump, or pumps, will operate correctly over the range of system conditions. Figure 6-19 shows two system curves and the pump operating range that each pump will experience for a system with three identical pumps in parallel. This plot of the overall pump curve with the two expected system curves illustrates how the flow rate and total discharge head is determined.

As mentioned earlier, the flow rate delivered by a system can be controlled through throttling (see Fig. 6-16). Control valves that can provide increased system friction losses as they are "throttled" or partially closed are utilized as they can effectively furnish control of the system. The degree of throttling can be adjusted as conditions warrant by adjustment of the opening of the valve (see chapter 8). For more permanent throttling, orifice plates may be installed in a system. While orifice plates are inexpensive and can provide effective throttling, they are not a readily modified form of throttling as the orifice plate must be changed (the orifice hole diameter). For a system with a pump, a bypass circuit can be employed to recirculate a portion of the flow from the pump discharge to the pump suction. However, these control strategies waste energy as energy contained in the flowing liquid is decreased by additional frictional losses.

Specific changes to the pumps can control system flow rate also. Pump impeller diameter and/or rotational velocity may be changed as discussed above. The pump

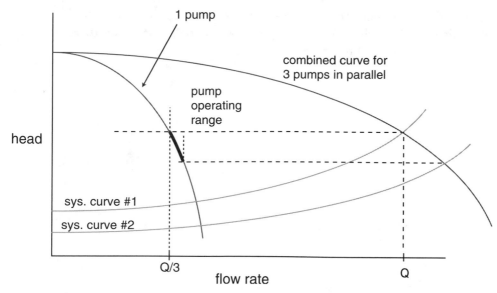

Figure 6-19. System operation for three identical pumps in parallel with changing system curve. The pump operating range is shown.

affinity laws predict how these changes affect the pump, and the operating point of the system can be determined from the intersection of the "new" pump curve and the system curve. And, the use of multiple pumps, a common design, allows some pumps to be turned on or off as necessary. Using Fig. 6-18 for example, for the lowest system demand, pump #1 could be in operation giving an operating point at the intersection of the system curve and the pump curve for pump #1. As the demand increases, pump #2 can be energized as well as pump #1 producing a greater discharge head and flow rate (at the "new" operating point defined by the intersection of the system curve and the combined curve). So the use of multiple staggered or booster pumps (in parallel or in series) and turning each on and off, can provide an increased operating range for the system, and provide some control.

Specific Speed

For pumps that are geometrically similar, dividing the square root of the capacity coefficient C_Q by the head coefficient C_H raised to the 3/4 power (which removes the impeller diameter) yields

$$\frac{C_Q^{1/2}}{C_H^{3/4}} = \frac{\left(\dfrac{Q}{nD^3}\right)^{1/2}}{\left(\dfrac{gH_t}{n^2D^2}\right)^{3/4}} \tag{6-35}$$

where n is the impeller rotational velocity, Q is the flow rate, and H_t is the total dynamic head. Canceling D from the numerator and denominator, dropping g (by convention), and simplifying yields

$$\frac{C_Q^{1/2}}{C_H^{3/4}} = \frac{nQ^{1/2}}{H_t^{3/4}} \tag{6-36}$$

This ratio is called the *specific speed* and is given by

$$n_s = \frac{6.67 \cdot nQ^{1/2}}{H_t^{3/4}} \tag{6-37}$$

where n is in revolutions per minute, Q is in cubic meters per minute, and H_t is in meters, or by

$$n_s = \frac{nQ^{1/2}}{H_t^{3/4}} \tag{6-38}$$

where n is in revolutions per minute, Q is in gallons per minute, and H_t is in feet.

The specific speed of the pump is a characteristic of the pump type, as shown in Fig. 6-20. The utility in calculating the anticipated (or required) pump specific speed is that it can assist in choosing the appropriate type of pump.

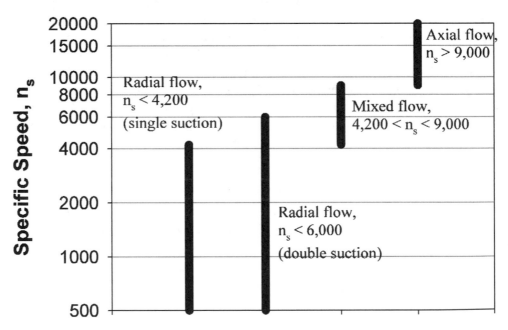

Figure 6-20. Specific speeds for centrifugal pumps. *Source:* Data from Hydraulic Institute (1975).

Specific Speed Example

A flow of 400 gpm must be pumped uphill for a total head (static and frictional) of 50 ft. The pump is driven by an electric motor at a speed of 1750 rpm. What type of centrifugal pump should be selected?

The specific speed is given by

$$n_s = \frac{nQ^{1/2}}{H_t^{3/4}} = \frac{1,750 \cdot (400)^{1/2}}{(50)^{3/4}} = 1,900$$

From Fig. 6-20, a specific speed of 1,900 falls in the range for a radial flow centrifugal pump with either single or double suction.

Cavitation

Most substances can exist in more than one phase in the range of conditions that may be encountered in fluid systems. Possible phases that may be present are

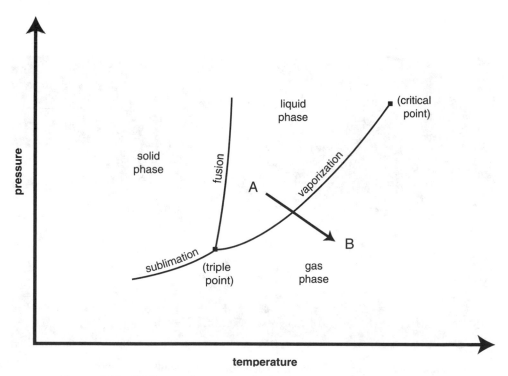

Figure 6-21. *PT* diagram for a general substance.

solids, liquids, and gases, as shown in Fig. 6-21. Conditions in a system may change from those where the substance exists as a liquid phase to those where the liquid volatilizes to a gas or vapor. This is illustrated by the arrow going from A (liquid phase) to B (gas phase) in Fig. 6-21. Within a piping system, the phase change from liquid to vapor phase is called cavitation. Sometimes cavitation can occur in the suction of a pump. Strictly speaking, cavitation occurs when the absolute pressure in the pump suction is below the fluid vapor pressure. The phenomenon occurs in this order:

1. Small bubbles form.
2. Water "boils," forming vapor bubbles.
3. Bubbles collapse in high-pressure areas (such as near pump impeller vanes).
4. Liquid rushes in on bubble collapse and can impact solid surfaces, causing damage.

Fluid rushing in to fill the voids left by the collapsing bubbles can damage nearby surfaces. An example of surface damage caused by cavitation is shown in Fig. 6-22. The degree of damage to the pump surfaces is dependent on the material subjected to cavitation. Some materials are able to withstand cavitation for somewhat longer periods of time before damage becomes significant. When subjected to cavitation damage, Monel (Ni-Cu alloy) has a relative life about twice that of mild steel, stainless steel about four times that of mild steel, titanium about six

Figure 6-22. Example of cavitation damage. *Source:* Sanks et al. (1998, p. 257), with permission from Elsevier.

times that of mild steel, and aluminum bronze about eight times that of mild steel (Hydraulic Institute, 2003).

Not only does the onset of cavitation produce damage to the pump, it also results in a drop in performance, as shown in Fig. 6-23. The pump curve that reaches the furthest to the right on the plot has no cavitation occurring. As the suction lift distance is increased (the vertical distance the pump must draw the liquid up to its suction), the tendency for the pressure in the suction line to drop below the vapor pressure is increased, and the pump curve drops off with a reduced discharge flow rate. The dashed arrow indicates the trend of reduced pump performance with increasing suction lift distance. As cavitation occurs, one observes a dramatic drop in delivery at a lower capacity than when the pump operates without cavitation. Manufacturers may define cavitation for a given pump as occurring when the total dynamic head decreases a certain percentage with increasing suction lift or resistance.

Net Positive Suction Head

A quantitative technique for determining the tendency for pump cavitation is the use of *net positive suction head* (NPSH) calculations. The NPSH *available* (NPSH$_A$) is

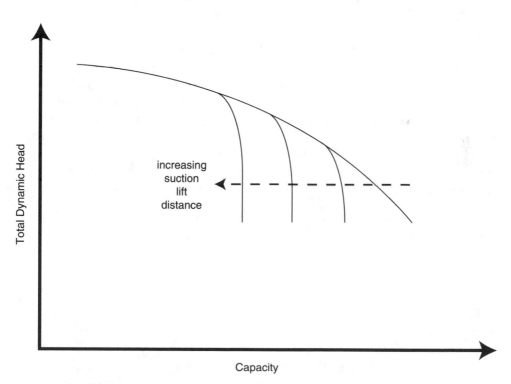

Figure 6-23. Typical pump performance degradation with increasing suction lift distance.

a measure of the absolute pressure in excess of the fluid vapor pressure in pump suction piping. It is calculated with (Sanks et al. 1998)

$$\text{NPSH}_A = h_{atm} + h_S - h_f - h_{sat} - FS \qquad (6\text{-}39)$$

where h_{atm} is the atmospheric pressure head, h_s is the static head above the eye of the pump impeller, h_f is the friction head loss in the pump suction, h_{sat} is the fluid vapor pressure in head units, and FS is a factor of safety. To prevent cavitation, engineers need to make sure that NPSH_A is greater than NPSH_R, the net positive suction head *required* by the pump to avoid cavitation. Cavitation may be determined in an operating system by "throttling" a pump suction line until the performance declines slightly. Values of NPSH_R for pumps are measured and reported by the pump manufacturers.

The factor of safety (FS) is used to account for

- uncertainties in NPSH_A calculations and
- variations in the reported pump NPSH_R owing to manufacturing variances and imperfections.

The actual value to use for a factor of safety is a matter of engineering judgment, but usual values range from 0.6 to 1.5 m (2 to 5 ft), or 20% to 35% of the NPSH_R (Sanks et al. 1998). The engineering team has to make the choice as to how conservative to be and should consult with the pump manufacturer.

Dissolved gases may also increase the tendency for cavitation above that for water without gases present. The cavitation potential for air and volatile compounds in water may be accounted for by including another term in Eq. 6-39, the partial pressure head of the dissolved gas, which will lower the NPSH_A slightly (up to 0.6 m) (Sanks et al., 1998). This is usually neglected but may be significant in some applications.

NPSH Example

Estimate the NPSH_A in the pumping system in Fig. 6-24. The suction lift (from the water surface in the suction-side tank to the pump suction (inlet) is 3.5 m. The friction loss in the suction line is 0.10 m. The following conditions apply: temperature = 15 °C, P_{atm} = 101.4 kPa, and P_{sat} = 4.14 kPa.

Solution

The pressures are

$$P_{atm} = 101.4 \text{ kPa} = 1.014 \times 10^5 \text{ kg/m·s}^2$$

$$P_{sat} = 1.709 \text{ kPa} = 1.709 \times 10^3 \text{ kg/m·s}^2$$

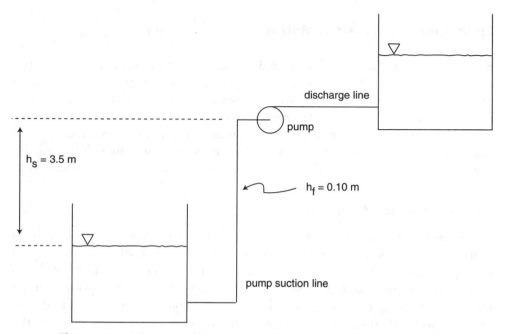

Figure 6-24. Schematic of the piping configuration for NPSH example.

Converting P_{atm} and P_{sat} to h_{atm} and h_{sat}, we get

$$h_{atm} = \frac{P_{atm}}{\rho \cdot g} = \frac{1.014 \times 10^5 \, \frac{kg}{m \cdot s^2}}{999.1 \frac{kg}{m^3} \cdot 9.81 \frac{m}{s^2}} = 10.3 \text{ m}$$

$$h_{sat} = \frac{P_{sat}}{\rho \cdot g} = \frac{1.709 \times 10^3 \, \frac{kg}{m \cdot s^2}}{999.1 \frac{kg}{m^3} \cdot 9.81 \frac{m}{s^2}} = 0.17 \text{ m}$$

Now the net positive suction head available is

$$\text{NPSH}_A = h_{atm} + h_s - h_f - h_{sat} - FS$$

and so

$$\text{NPSH}_A \text{ (without FS)} = 10.3 \text{ m} - 3.5 \text{ m} - 0.10 \text{ m} - 0.17 \text{ m} - 0.0 = 6.53 \text{ m}$$

If we choose $FS = 1$ m,

$$\text{NPSH}_A \text{ (without FS)} = 10.3 \text{ m} - 3.5 \text{ m} - 0.10 \text{ m} - 0.17 \text{ m} - 1.0 = 5.53 \text{ m}$$

Therefore a pump must be chosen with $\text{NPSH}_R < 5.53$ m.

Fundamentals of Electric Motors

Electric motors are commonly used to drive pumps in treatment systems. Electric motors are compact and versatile and require little maintenance. They can be obtained in a range of sizes with various power ratings. Drawbacks to the use of electric motors are that they (1) can be damaged by water from flood events, (2) are affected by electrical surges and outages, and (3) often require high starting currents. There are many types of electric motors used, including squirrel-cage induction motors, wound-rotor induction motors, and synchronous motors.

Squirrel-Cage Induction Motors

The squirrel-cage induction motor is the most common motor used for driving pumps because it is simple, requires little maintenance, is reliable and tough, and has high efficiency (Sanks et al. 1998). It uses alternating current and there are no connections or contacts to the rotor from the stator (the stationary portion of the motor surrounding the rotor). See Fig. 6-25. The rotor is composed of laminations of steel with slots for conductors (with the slots being insulated from the steel). Copper or aluminum rotor bars are used for the conductors. The rotational speed of the shaft is relatively constant, usually within a few percent difference in rotational speed from full load to no load. However, induction motors have high starting currents, typically around 600% of full load current (Sanks et al. 1998). The effect of the high starting current can sometimes be somewhat offset by using

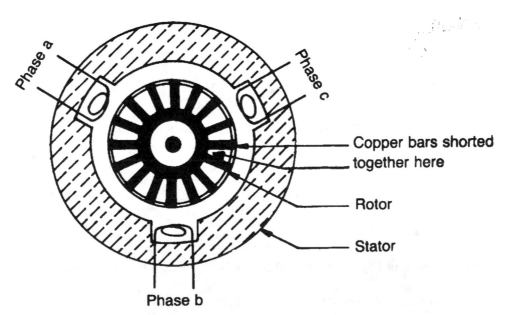

Figure 6-25. Schematic of a squirrel-cage electric motor. *Source:* Sanks et al. (1998, p. 406), with permission from Elsevier.

reduced voltage starters. Efficiencies of squirrel-cage induction motors vary from approximately 85% for a 4-kW (5-hp) motor to approximately 91% for a 75-kW (100-hp) motor (Sanks et al. 1998). There are higher efficiency motors available with advanced designs (thinner laminations in internal components, differently shaped rotors, etc.).

Wound-Rotor Induction Motors

The wound-rotor induction motor is not as commonly used as the squirrel-cage induction motor, but it is suited for conditions where there is a need for high starting torque. The rotor for a wound-rotor motor has windings in rotor slots, and these windings are connected via slip rings (contacts) to connections on the stator, as shown in Fig. 6-26. By varying the resistance of the windings on the rotor with switches, the startup current can be regulated to reduce the "inrush" current to approximately 125% of the full-load current (Sanks et al. 1998). It is now uncommon to use a wound-rotor induction motor, but they are still applicable when large electric motors [greater than 750 kW ($\approx$1000 hp)] are required (Sanks et al. 1998).

Synchronous Motors

Synchronous motors have a rotor with poles excited by direct current (DC) that produces a rotating field. See the schematic for a synchronous motor in Fig. 6-27. The speed can be regulated with the external magnetic flux from windings in the stator. As for wound-rotor motors, connections or contacts are necessary between

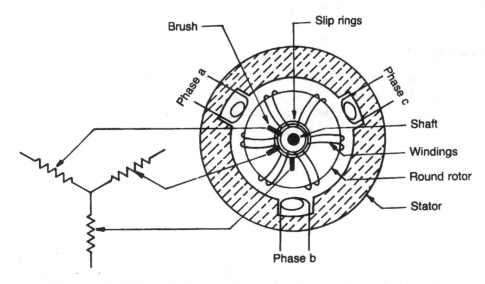

Figure 6-26. Schematic of a wound-rotor electric motor. *Source:* Sanks et al. (1998, p. 407), with permission from Elsevier.

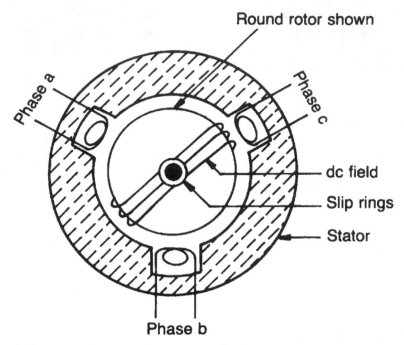

Figure 6-27. Schematic of a synchronous electric motor. *Source:* Sanks et al. (1998, p. 408), with permission from Elsevier.

the rotor and stator, in this case to supply the direct current to the rotor (but there are "brushless" motors without the contacts). Because of the high prices for these types of motors, they are not commonly used except for large power requirements [greater than 400 kW (≈500 hp)] (Sanks et al. 1998).

Motor Speed

The speed, or rotational velocity at which a motor rotor operates, depends on the rotational velocity of the magnetic field created by the stator. The stator can have a number of magnetic "poles." The rotational velocity of a rotor, or its synchronous speed, in radians per second is calculated from (Sanks et al. 1998)

$$\omega\left(\text{in}\ \frac{\text{radians}}{\text{s}}\right) = \frac{2\pi \cdot \text{Hz}}{n} \tag{6-40}$$

or from

$$\omega\left(\text{in}\ \frac{\text{revolutions}}{\text{min}}\right) = \frac{120 \cdot \text{Hz}}{n} \tag{6-41}$$

where n is the number of poles.

So the operational rotational velocity is dependent on the driving electrical frequency and the number of poles. A 60-Hz alternating current power supply rotates 360 degrees (one revolution around the stator), for a one-pair pole motor (north and south poles; $n = 2$). So the one-pair pole motor would rotate at

$$\omega = \frac{120 \cdot Hz}{n} = \frac{120 \cdot 60}{2} = 3,600 \text{ rpm}$$

It is common to use 900, 1,200, or 1,800 rpm electric motors for water, wastewater, and sludge pumping (Sanks et al. 1998).

Electric Motor Torque

The power exerted by a pump is the rate of energy transfer to the fluid at constant conditions (speed and flow rate). Upon startup of a pump, the motor needs to be able to accelerate the impeller and fluid, and this requires torque. The torque needed to accelerate the pump from completely stopped to its operating speed is a characteristic of the pump and this information is supplied by the pump manufacturer. The National Electrical Manufacturers Association (NEMA) and the International Electrotechnical Commission (IEC) classify electric motors by "designs" that have similar characteristics. NEMA Design A and B and IEC Design N are applied when the starting torque is low as it is for many centrifugal pumps. NEMA Design C and IEC Design H are for applications where the starting torque may be greater than expected at operating speed and greater than Designs A, B, and N are capable of. NEMA Design D is employed where high starting torques are needed, or when frequent starting of the motor occurs. However, Design D typically has a greater slip and therefore lower efficiency than the other designs. See Fig. 6-28 for a comparison of the torques produced by electric motors. For the motor and pump to accelerate to operating speed quickly, the torque produced by the motor should be greater than the torque required by the pump (available from the pump manufacturer) at all speeds.

Motor Features

The shafts for electric motors are typically solid except for deep well pumps with vertical shafts, which may be hollow. These shafts must be supported by bearings, which are commonly antifriction bearings (roller bearings). Sleeve bearings may be used in some large motors.

Motors may be damaged by high winding temperatures. Consequently, prevention of excessive temperatures is desirable to avoid permanent damage to the motor. Temperature sensors provide a defense against high temperatures by deenergizing the motor and/or alerting the operators during temperature extremes. There are three main types of temperature sensors:

- Bimetallic element—A bimetallic element is a simple direct-acting apparatus that can open or close as temperature rises above a threshold value consid-

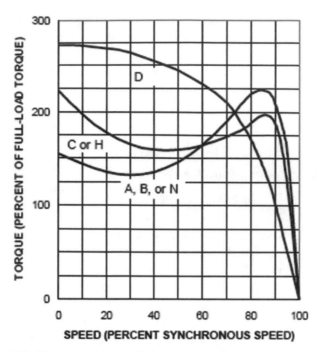

Figure 6-28. Torque versus speed characteristics for NEMA and IEC electric motors. *Source:* NEMA (2001), with permission from the National Electric Manufacturers Association.

ered dangerous for the hardware. The contacts control the operating circuit for the motor and can reset automatically with a temperature drop into the proper operating range.

- PTC thermistor—A positive temperature coefficient (PTC) thermistor is a small temperature sensor that can be located adjacent to the pump windings. The resistance of the PTC thermistor is low until a threshold temperature is reached, at which point the resistance rises very quickly as a function of temperature. It can be connected to a control circuit to monitor the resistance and deenergize the pump when the resistance (temperature) increases above a set limit.
- RTD—A resistance temperature detector (RTD) provides a relatively linear resistance as a function of temperature. The resistance of the RTD can be continuously measured and converted to temperature.

Despite their high cost, vibration monitors are sometimes used as part of a preventative maintenance protocol. Vibration responses from sensors located on the motor are collected and analyzed on a periodic basis. Changes in the signals may indicate a possible future failure (such as vibration from a failing roller bearing).

Moisture sensors are sometimes used in motors on submersible pumps (Sanks et al. 1998). The sensors are capacitance-type probes that indicate when water passes by seals, as a result of the increased conductivity of water. Moisture sensors are located between the outer and inner seals on a submersible motor.

Motor Efficiency

As discussed earlier, the power delivered to fluid by a pump is less than the power delivered to the pump input shaft. There are also power losses through various components of the electric motor. Figure 6-29 provides a typical breakdown of the power losses expected in an electric motor. Because of these losses, the power delivered to an electrical motor output shaft is less than the electrical power supplied to the motor. The efficiency of an electrical motor is

$$E_M = \frac{P_{shaft}}{P_{electrical}}$$ (6-42)

Just as the efficiencies of pumps are reported by pump manufacturers, the efficiencies of electrical motors are available from the motor manufacturers. The efficiency of a pump was defined as the power delivered to the fluid divided by the power supplied to the input shaft of the pump. The *overall* efficiency of a pump driven by an electrical motor is

$$E_{overall} = E_P \cdot E_M$$ (6-43)

which in terms of power is

$$E_{overall} = \frac{P}{P_{shaft}} \cdot \frac{P_{shaft}}{P_{electrical}} = \frac{P}{P_{electrical}}$$ (6-44)

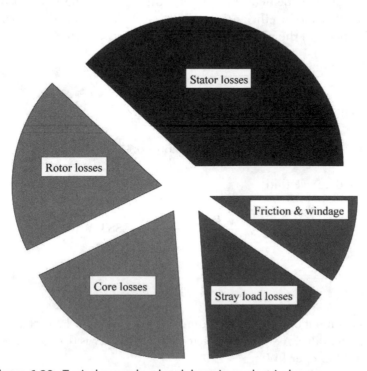

Figure 6-29. Typical power loss breakdown in an electrical motor.

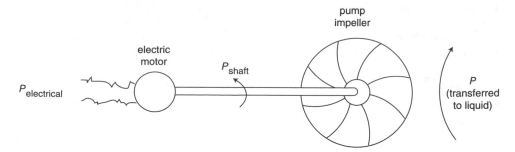

Figure 6-30. P, P_{shaft}, and $P_{electrical}$ for an electrical motor coupled to a pump.

P, P_{shaft}, and $P_{electrical}$ are depicted in Fig. 6-30 for an electric motor coupled to a pump. Through determination of the operating point of a pumping system, the flow rate and total dynamic head produced by the pump are fixed. From the efficiencies of the pump and motor, $P_{electrical}$ can be found and the electrical operating cost determined.

Example

If the power that must be delivered to water during pumping is 18.5 kW, and the pump and electric motor efficiencies are 60% and 35%, respectively, what are the overall efficiency and the required power of the electric motor?

Solution

The overall efficiency is

$$E_{overall} = E_P \cdot E_M = 0.60 \cdot 0.35 = 0.21$$

From Eq. 6-44, we find

$$P_{electrical} = \frac{P}{E_{overall}} = \frac{18.5 \text{ kW}}{0.21} = 88 \text{ kW}$$

Example

Calculate the annual energy cost to pump 12 ft³/s though a 12 in. ball valve, gate valve, and globe valve. Assume that the overall efficiency is 0.72 and the electrical power cost is \$0.20 per kW·hr.

Solution

From Fig. 5-15, K is 0.04 for the ball valve, 0.1 for the gate valve, and 4.5 for the globe valve. The fluid velocity through the valves is 15.3 ft/s (flow rate divided by the flow area). From Eq. 5-73 for the ball valve:

$$h_f = \left(4\frac{f\Delta x_{total}}{D} + \sum_{\substack{\text{all valves} \\ \text{\& fittings}}} K\right)\left(\frac{V^2}{2g}\right) = (0+0.04)\left(\frac{\left(15.3\frac{ft}{s}\right)^2}{2\cdot 32.2\frac{ft}{s^2}}\right) = 0.145\ \text{ft}$$

Similarly, the head loss through the gate valve is 0.363 ft and through the globe valve is 16.3 ft. In this example, $H_t = h_f$. From Eq. 6-10 for the ball valve, the power required to be transferred to the fluid can be calculated:

$$P = g\cdot\rho\cdot Q\cdot H_t = 32.2\frac{ft}{s^2}\cdot 62.367\frac{lbm}{ft^3}\cdot 12\frac{ft^3}{s}\cdot 0.145\ \text{ft}\cdot\frac{lbf\cdot s^2}{32.2\ ft\cdot lbm} = 109\frac{ft\cdot lbf}{s}$$

Similarly, P is 272 ft·lbf/s for the gate valve and 12,200 ft·lbf/s for the globe valve.

The electrical power needed for the ball valve is:

$$P_{electrical} = \frac{P}{E_{overall}} = \frac{109\frac{ft\cdot lbf}{s}}{0.72}\left(\frac{0.7457\ kW}{550\frac{ft\cdot lbf}{s}}\right) = 0.20\ kW$$

Similarly, 0.51 kW is required for the gate valve and 23.0 kW for the globe valve.

Therefore the annual electrical cost for the ball valve is:

$$0.20\ kW \cdot \$0.20\frac{1}{kW\cdot hr}\cdot 8,760\frac{hr}{yr} = \$350\ \text{per year}$$

And the annual electrical cost is $900 for the gate valve, and $40,000 for the globe valve.

Variable-Speed Drives

Although the least complex motor and pump installation is one that operates at one speed, a single-speed pump may not be the most energy efficient over wide-ranging conditions. To control a single-speed pump system, a control valve may be throttled, as shown in Fig. 6-16 and discussed earlier. This produces an increased friction loss that results in an increased energy loss by the system and the pump. This is an energy consumption that is greater than required. More sophisticated variable-speed drives (VSDs) provide for variable pump performance that can change as needed by the system. We know from the pump affinity laws that changing the impeller speed will affect pump capacity, total dynamic head, and power (Eqs. 6-23, 6-24, and 6-25). By providing variable impeller speed, VSDs may reduce energy utilization, lower maintenance costs, and increase reliability. High starting currents can be reduced or eliminated when VSDs replace multiple pumps, as less pump starting may result. By eliminating the need for throttling to control the system, or at least reducing the amount of throttling needed, VSDs can reduce the high-pressure drops from the throttling devices that can cause noise, erosion, or cavitation in the system. VSDs also usually produce better overall control of the system than through throttling system control. VSDs may be used to control the discharge pressure of the pump, the differential pressure across the pump (from suction to discharge), flow rate delivered by the system, static liquid height in a supply or discharge tank, or power developed by the motor.

The speed that a pump operates may be varied mechanically or electrically. Mechanical variation consists of variable-speed pulley and belt combinations, hydrokinetic fluid couplings, and eddy current couplings. Mechanical variation systems may not have high drive efficiencies in all speed ranges and can produce heat output (energy loss). The most common electrical VSD is through pulse-width modulation, and is commonly known as variable-frequency drive. The speed of the electrical motor is proportional to the frequency provided to the motor by the variable-frequency drive. Variable-frequency drives are capable of a wide range of speeds, from 2% to 200% (Europump and Hydraulic Institute 2004), with high efficiencies. Other electrical speed controllers used less often than pulse-width modulation are pulse-amplitude modulation and current source inverters. See Sanks et al. (1998) and Europump and Hydraulic Institute (2004) for more information on variable-speed drives.

Symbol List

C_H	head coefficient
C_P	power coefficient
C_Q	capacity coefficient
D	impeller diameter
E_M	motor efficiency
E_P	pump efficiency
FS	factor of safety for NPSH calculations
h	fluid head
h_{atm}	atmospheric pressure head
h_f	friction head loss in pump suction

h_s	static head above eye of pump impeller
h_{sat}	fluid vapor pressure in head units H head delivered by pump
H_t	total dynamic head
$\dot{m}$	mass flow rate
n	rotational velocity; number of poles in an electric motor
$NPSH_A$	net positive suction head available
$NPSH_R$	net positive suction head required
P	power, power delivered to the liquid, pressure
P_{sat}	vapor pressure
$P_{electrical}$	input electrical power
P_{shaft}	power delivered to the pump shaft
Q	volumetric fluid flow rate
r_1	inner radius of impeller
r_2	outer radius of impeller
SG	specific gravity of fluid
T	torque on impeller
V_1	absolute fluid velocity at inner radius of impeller
V_2	absolute fluid velocity at outer radius of impeller
α_1	angle of V_1 to tangent of inner radius of impeller
α_2	angle of V_2 to tangent of outer radius of impeller
γ	fluid specific weight
ω	rotational velocity
ρ	fluid density

Problems

1. What power must be transferred to water to pump it against 2.2 ft of friction head at 500 gpm with a static height increase of 80 ft? If the pump efficiency is 75% and the motor efficiency is 70%, what horsepower electric motor is necessary?

2. Water at 50 °F is pumped at 100 gpm through a 2-in. inner diameter commercial steel suction line that is 25 ft long. The suction lift is 15 ft. The suction line has a square-edged inlet, four regular flanged 90° elbows, and a ball valve. What must the pump $NPSH_R$ be?

3. What type of centrifugal pump should be selected that is to be driven at 3600 rpm, pumping 3.0 m³/min, against a head of 7.5 m?

4. A pump with the $7^7/_{16}$-in. diameter impeller, whose characteristic curve is shown in Fig. 6-15, is being used to transfer 12 °C water to a storage tank. The pump intermittently runs during the day, transferring a total of 50 m³ per day. If the total dynamic head is 43 m, and the electric motor efficiency is 55%, how much electrical energy is expended each day in kW-hr? If corrosion increases the total dynamic head to 49 m, how much electrical energy is expended? What percent increase in electrical cost would be expected?

References

Europump and Hydraulic Institute (2004). *Variable Speed Pumping, A Guide to Successful Applications*, Elsevier, Oxford.

Hydraulic Institute (1975). *Standards for Centrifugal, Rotary & Reciprocating Pumps*, Hydraulic Institute, Parsippany, NJ.

Hydraulic Institute (2003), *Centrifugal Pumps: Fundamentals, Design, and Applications*, E Learning CD, Hydraulic Institute, Parsippany, NJ.

Karassik, Igor J. (1981). *Centrifugal Pump Clinic*, Marcel Dekker, New York.

NEMA (2001). *NEMA Standards Publication MG 10–2001, Energy Management Guide for Selection and Use of Fixed Frequency Medium AC Squirrel-Cage Polyphase Induction Motors*, National Electric Manufacturers Association, Rosslyn, VA.

Sanks, R. L., Tchobanoglous, G., Bosserman, B. E., II, and Jones, G. M., eds. (1998). *Pumping Station Design*, Butterworth-Heinemann, Boston, MA.

CHAPTER 7

Friction Loss in Flow through Granular Media

Chapter Objectives

1. Quantify friction loss in fluid flow through granular media.
2. Estimate minimum fluidization velocity for packed beds.

Granular Media Used in Treatment Systems

Liquid flow through solid granular media is a common process in many treatment systems. Granular media are used in sand and mixed media filtration, granular activated carbon contactors, and ion exchange processes. As the fluid must move in the stationary pores of the media, energy is lost from skin friction (fluid flowing past granular surfaces) and form friction (fluid flowing around grains). Engineers must be able to predict the frictional loss to properly size the pumps and piping servicing the process employing granular media.

Granular media processes may be contained by closed vessels, as shown by the ion exchange process in Fig. 7-1, or open vessels, as for the slow sand filter in Fig. 7-2. In an open vessel, the surface of the media or the fluid above the media is open to the atmosphere. In a closed vessel, the difference in pressure between the inlet and discharge sides of the media must be accounted for in the system design, or the desired flow rate may not be attainable. In the open vessel, the desired flow must be achievable with the head of fluid above the media in a downflow configuration.

Granular Media Filtration

Filtration is the removal of suspended solids and colloidally sized particles (particles with one dimension between 1 nm and 1 μm) by passing through a media such as sand. The particles are destabilized and will attach to the media surfaces in the filter if the conditions are amenable to attachment. See Fig. 7-3. The removal of particles from the liquid occurs within the media bed, but the flow through the filter produces friction loss as well. As solids are removed from the

Figure 7-1. Closed-vessel ion-exchange process.

Figure 7-2. Slow sand filter, open-vessel filter process. Above the media is open to atmospheric pressure.

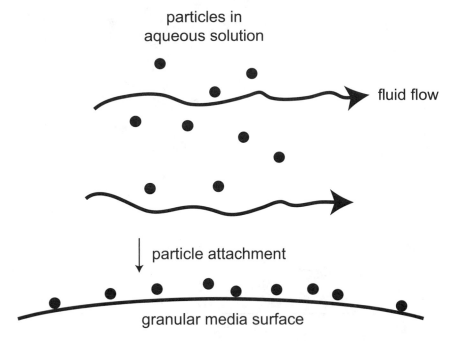

Figure 7-3. Attachment during granular media filtration process.

fluid stream, the pore volume decreases and the frictional loss increases above the clean bed friction loss, as illustrated in Fig. 7-4. As available surface area becomes occupied by attached solids during a filter run, the quality of the effluent decreases. Types of granular media filters include the following:

- Rapid, granular media filters: These filters may be single media (sand), dual media (sand and anthracite), or triple media (sand, anthracite, and garnet). Filtration rates usually range between 80 and 400 $L/m^2 \cdot min$ (≈ 2 to 10 gpm/ft^2) (Tchobanoglous and Schroeder 1985), with media depths less than 1 m (≈ 3 ft). The sand filter must be backwashed when the effluent no longer meets acceptable quality levels, or when the frictional loss is too high for the system.
- Slow sand filters: Slow sand filters have a much lower filtration rate than rapid sand filters, with filtration rates ranging from 2 to 5 $L/m^2 \cdot min$ (≈ 0.05 to 0.12 gpm/ft^2) (Tchobanoglous and Schroeder 1985). In these filters, suspended solids and colloids are removed in a surface mat, called the "schmutzdecke". When the slow sand filter becomes clogged (manifested by high head loss), the upper layer is removed and replaced with clean sand.

Granular Activated Carbon Contactors

Granular activated carbon (GAC) is the granular form of activated carbon, which is produced by heating an organic such as hardwood, coal, or nut shells without

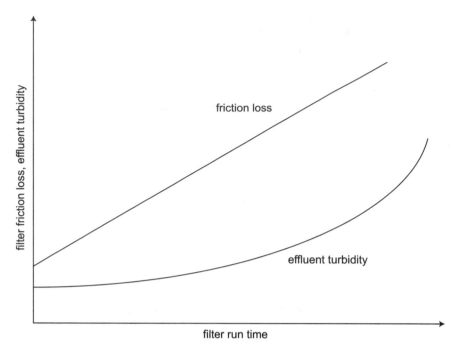

Figure 7-4. Normal operation of a sand filter showing increasing friction loss with run time and increase in effluent turbidity.

oxygen present (pyrolysis). The carbon has a high surface area, on the order of 1,000 m^2/g. Organic contaminants (e.g., benzene and toluene) have an affinity for the activated carbon and so are removed from the fluid stream as it passes through the activated carbon. Granular activated carbon may be used as fixed beds in packed bed, closed vessel contactors similar to that shown in Fig. 7-1. It can also be used in open vessel contactors. As the fluid passes through the GAC bed, it loses energy from frictional loss.

It is common for more than one fixed bed contactor to be operated in series, one after another, so that the effluent quality from the series can be maintained. When breakthrough of contaminant occurs in the effluent from the first contactor, it can be taken offline, and the GAC changed (regenerated or disposed of). When two contactors in series are used in this fashion, the second contactor prevents contaminants from being discharged because the second contactor's carbon still effectively removes contaminants after the first contactor's carbon is saturated with contaminants. See Fig. 7-5.

Friction Loss in Granular Media

Fluids flowing through beds of granular media will lose energy. When designing a system using a granular media process, it is important to be able to predict the

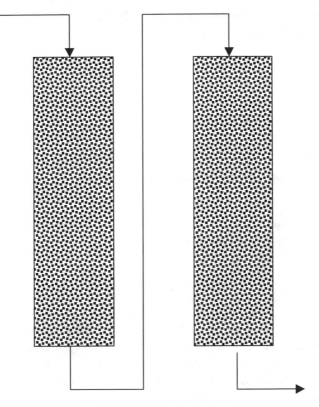

Figure 7-5. Series configuration fixed-bed granular activated carbon contactors in operation.

frictional loss for flow through these packed beds. The following presents a derivation for an equation that may be used to calculate friction loss for fluids flowing through packed beds.

Characteristics of Fluid Flow in Granular Media

For a packed bed of granular media, the media grains are supported on one another, and the fluid must flow around the grains to pass through the bed. The fluid cannot flow through the solid, so the solid portion of the bed (made up of the grains) is precluded from fluid flow. The void volume fraction available for fluid flow, defined as *porosity* ε, is

$$\varepsilon = \frac{\text{volume of voids in bed}}{\text{total bed volume}} \tag{7-1}$$

Typical porosity values are on the order of 0.3 to 0.4 dependent on the grain size, shape, and packing. The presence of the solid grains in the contactor increases the fluid velocity through the contactor relative to the velocity in a contactor that is empty because the cross-sectional area for flow is reduced by the solid grains. The empty bed velocity V_0 is also called the superficial velocity. The velocity in the pores that the fluid flows through is

$$V_{pore} = \frac{V_0}{\varepsilon} \tag{7-2}$$

For fluid flow through granular media, we can characterize the flow with the Reynolds number as was done for conduit flow. The Reynolds number *in general terms* is written for a characteristic length, characteristic velocity, fluid density, and fluid viscosity as

$$\Re = \frac{\text{characteristic length} \cdot \text{characteristic velocity} \cdot \text{fluid density}}{\text{absolute fluid viscosity}} \tag{7-3}$$

For flow through a packed bed, it is common, but not universal, to choose the grain diameter d as the characteristic length and the fluid superficial velocity V_0 as the characteristic velocity, so that

$$\Re = \frac{d \cdot V_0 \cdot \rho}{\mu} \tag{7-4}$$

where ρ is the fluid density and μ is the absolute fluid viscosity.

The value of the Reynolds number, as for flow through a pipe, provides an indication of the type of flow in a packed bed. For flow through granular media with low Reynolds numbers, $\Re \ll 1$, laminar flow is expected. For low-Reynolds-number flow, viscous effects dominate and there is little lateral mixing during flow through the media. There are no cross-currents or eddies. For flow with high Reynolds numbers, $\Re \gg 1$, turbulent flow is expected as viscous effects are negligible. Random fluid motion predominates in that eddies form and lateral mixing occurs.

Derivation of Friction Loss Equation

Consider a cylindrical element with the axis running parallel to flow in a packed bed as illustrated in Fig. 7-6. For this derivation, assume that steady-state conditions exist and that the fluid is incompressible (constant ρ). A fluid flowing through the cylinder will exert a force on the solid surfaces because of wall and form drag. That is, the fluid will lose energy. We can define a dimensionless friction factor e, such that the force exerted on the surfaces by the fluid is a function of the kinetic energy per unit volume of the fluid:

$$F_K = A \cdot K \cdot e \tag{7-5}$$

where A is the wetted surface area of the cylinder, K is the kinetic energy of the fluid per unit volume ($\frac{1}{2} \cdot \rho \cdot V_{pore}^2$) and e is a dimensionless friction factor. Conducting a force balance on the fluid in the element, as shown in Fig. 7-7, we have

$$F_0 - F_K - F_L = 0 \tag{7-6}$$

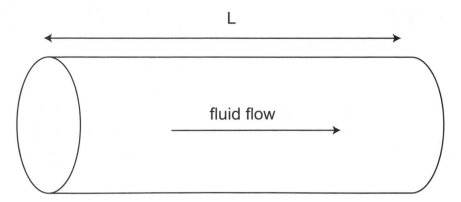

Figure 7-6. Cylinder representing element for fluid flow in granular media.

where F_0 is the force on the upstream face of the cylinder due to the upstream pressure and F_L is the force on the downstream face due to the downstream pressure.

Since the force exerted by pressure is the pressure multiplied by the area (πr^2 in this case), the force balance becomes

$$F_K = F_0 - F_L = P_0 \cdot \pi \cdot r^2 - P_L \cdot \pi \cdot r^2 \qquad (7\text{-}7)$$

Simplifying this expression gives

$$F_K = (P_0 - P_L) \cdot \pi \cdot r^2 = \Delta P \cdot \pi \cdot r^2 \qquad (7\text{-}8)$$

Combining Eqs. 7-8 and 7-5 yields

$$(2\pi \cdot r \cdot L)(\tfrac{1}{2} \cdot \rho \cdot V_{\text{pore}}^2) \cdot e = \Delta P \cdot \pi \cdot r^2 \qquad (7\text{-}9)$$

where L is the cylinder length. The friction loss in the cylinder is ΔP. Substituting in the expression $\Delta P = \rho \cdot g \cdot h$ gives

$$(2\pi \cdot r \cdot L)(\tfrac{1}{2} \cdot \rho \cdot V_{\text{pore}}^2) \cdot e = \rho \cdot g \cdot h \cdot \pi \cdot r^2 \qquad (7\text{-}10)$$

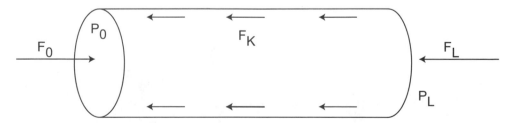

Figure 7-7. Force balance on cylinder representing element in granular media.

Rearranging to solve for the friction loss in head units, we get

$$h = e \frac{L}{r} \frac{V_{pore}^2}{g} \quad (7\text{-}11)$$

For fixed beds of granular media, the area for flow is not the full, complete tubular area. Because the pores through which the fluid flows are not of circular cross section, the hydraulic radius must be used, so r is changed to r_H. Substituting for V_{pore} from Eq. 7-2 gives

$$h = e \frac{L}{r_H} \frac{\left(\dfrac{V_0}{\varepsilon}\right)^2}{g} \quad (7\text{-}12)$$

The hydraulic radius of the cylinder can be expressed as

$$r_H = \frac{\text{cross section available for flow}}{\text{wetted perimeter}} = \frac{\text{volume available for flow}}{\text{total wetted surface area}} \quad (7\text{-}13)$$

Dividing the numerator and denominator by the bed volume gives

$$r_H = \frac{\text{volume of voids/bed volume}}{\text{wetted surface area/bed volume}} = \frac{\varepsilon}{a_v(1-\varepsilon)} \quad (7\text{-}14)$$

where a_v is the specific surface area, which is defined as

$$a_v = \frac{\text{total grain surface area}}{\text{volume of grains}} = \frac{\pi d^2}{\frac{1}{6}\pi d^3} = \frac{6}{d} \quad (7\text{-}15)$$

Substituting Eq. 7-15 into Eq. 7-14 for a_v gives an equation for the hydraulic radius for flow through a packed bed of granular media:

$$r_H = \frac{\varepsilon}{\frac{6}{d}(1-\varepsilon)} \quad (7\text{-}16)$$

Substituting Eq. 7-16 into Eq. 7-12, we have for the head loss equation

$$h = 6e \frac{(1-\varepsilon)}{\varepsilon^3} \frac{L}{d} \frac{V_0^2}{g} \quad (7\text{-}17)$$

By defining a "new" friction factor, f, as $6 \cdot e$, Eq. 7-17 becomes

$$h = f \frac{(1-\varepsilon)}{\varepsilon^3} \frac{L}{d} \frac{V_0^2}{g} \quad (7\text{-}18)$$

Grains that make up a packed bed may not be completely spherical, and the sphericity may be expected to affect the frictional loss. The head loss can be adjusted for sphericity of the grains in the bed with a sphericity or shape factor, ϕ:

$$h = \frac{f}{\phi} \frac{(1-\varepsilon)}{\varepsilon^3} \frac{L}{d} \frac{V_0^2}{g} \tag{7-19}$$

This is an equation to calculate the clean-bed head loss in granular media and is called the Kozeny-Carman or *Ergun* equation (see Droste 1997). The Kozeny-Carman friction factor is defined as

$$f = 150\left(\frac{1-\varepsilon}{\Re}\right) + 1.75 \tag{7-20}$$

where $\Re$, the Reynolds number, is quantified with

$$\Re = \frac{d \cdot V_0 \cdot \rho}{\mu} \tag{7-21}$$

The Kozeny-Carman friction factor, Eq. 7-20, comprises two terms, a laminar term and a turbulent term. The laminar term dominates at low Reynolds numbers (approximately <1), and the turbulent term dominates at high Reynolds numbers (approximately >1000). See Fig. 7-8.

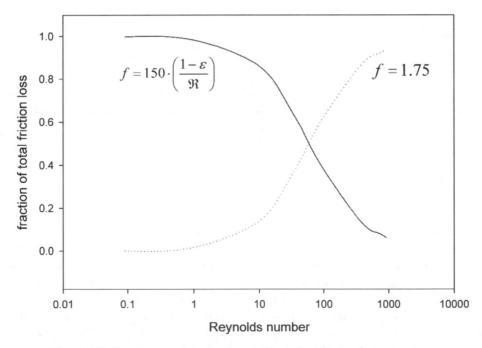

Figure 7-8. Dominance of the laminar and turbulent friction factor terms.

Example

Water at 20 °C is passing through a 1.0-m deep sand filter in a 1.5-m-diameter cylinder. The filtration rate is 125 L/m²·min and the grain size is 0.50 mm. Assume the porosity is 0.38 and the sphericity is 0.9. What is the head loss?

Solution

Since 125 L/m²·min = 2.083×10^{-3} m/s, the empty bed or superficial velocity is 2.083×10^{-3} m/s. At 20 °C, the fluid density is 998.21 kg/m³, and the absolute viscosity is 1.002 cP or 1.002×10^{-3} kg/m·s.

Calculating the Reynolds number we have

$$\Re = \frac{d \cdot V_0 \cdot \rho}{\mu} = \frac{(0.5 \times 10^{-3}\,\text{m}) \cdot \left(2.083 \times 10^{-3}\,\dfrac{\text{m}}{\text{s}}\right) \cdot 998.21\,\dfrac{\text{kg}}{\text{m}^3}}{1.002 \times 10^{-3}\,\dfrac{\text{kg}}{\text{m} \cdot \text{s}}} = 1.04$$

and so the Kozeny-Carman friction factor is

$$f = 150\left(\frac{1-\varepsilon}{\Re}\right) + 1.75 = 150\left(\frac{1-0.38}{1.04}\right) + 1.75 = 91.2$$

The Kozeny-Carman equation gives

$$h = \frac{f}{\phi}\frac{(1-\varepsilon)}{\varepsilon^3}\frac{L}{d}\frac{V_0^2}{g} = \frac{91.2}{0.9}\frac{(1-0.38)}{0.38^3}\frac{1\,\text{m}}{0.5 \times 10^{-3}\,\text{m}}\frac{\left(2.083 \times 10^{-3}\,\dfrac{\text{m}}{\text{s}}\right)^2}{9.81\,\dfrac{\text{m}}{\text{s}^2}} = 1.0\,\text{m}$$

Thus the friction loss through the sand filter is 1.0 m.

Fluidization of Granular Media

When the head loss or water quality in the effluent from a sand or mixed media filter is no longer acceptable, the media is backwashed. In backwashing, the fluid flow is reversed through the bed, and fluid is pumped upward to fluidize and expand the bed. The high shear forces and grain-to-grain abrasion produced during fluidization detaches captured particles from the media, allowing the particles to be removed from the bed with the fluid through the top of the filter.

Fluidization of a bed of granular media is produced as the upflow fluid velocity is increased through the bed until the bed starts to expand. As the velocity increases, the drag on individual grains increases and grains move apart and become suspended in the flow. Once the grains start to separate from each other (becoming suspended in the flow), the bed acts like a fluid in that it can be pumped and poured like a fluid, giving rise to the term *fluidization*.

The condition for fluidization is that the pressure drop through the bed counterbalances the weight of the bed per unit cross section. That condition defines the minimum fluidization velocity for the fluid in the bed. For fluidization, the force balance on a bed of granular media is shown in Fig. 7-9. The force balance is

$$F_0 - F_L - F_g = 0 \qquad (7\text{-}22)$$

where F_g is the force of gravity on the bed. $F_0 - F_L$ is the difference in the pressure force from the upstream face to the downstream face, which is due to the pressure loss from the fluid flowing through the bed. Force is equal to pressure P multiplied by area A, and we can substitute for the net force of gravity of the bed to get

$$\Delta P \cdot A = \frac{g}{g_c}(1 - \varepsilon)(\rho_g - \rho)\mathbf{v}_{\text{bed}} \qquad (7\text{-}23)$$

where ρ_g is the grain density. Because $\Delta P = \rho g h$ and $\mathbf{v}_{\text{bed}} = LA$, we have

$$\rho g h A = \frac{g}{g_c}(1 - \varepsilon)(\rho_g - \rho)LA \qquad (7\text{-}24)$$

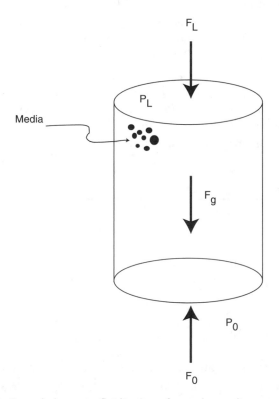

Figure 7-9. Force balance on fluidization of granular media.

Dividing by A gives

$$\rho g h = \frac{g}{g_c}(1-\varepsilon)(\rho_g - \rho)L \tag{7-25}$$

Equation 7-25, obtained from a force balance, can be solved simultaneously with the Kozeny-Carman equation, Eq. 7-19 (with Eqs. 7-20 and 7-21), to find the fluidization velocity, $V_{fluidization}$. It is not possible to arrange an equation explicit in $V_{fluidization}$ from these equations, but they may be solved with various techniques/software packages (e.g., MathCad®, Mathematica®, Microsoft Excel®). However, simplifications may be made in the two extreme flow regimes, when completely laminar or completely turbulent:

$$V_{fluidization} = \frac{g(\rho_g - \rho)}{150\mu}\frac{\varepsilon^3}{1-\varepsilon}\phi^2 d^2 \qquad (\text{for } \Re < 1) \tag{7-26}$$

$$V_{fluidization} = \left[\frac{\phi d g(\rho_g - \rho)\varepsilon^3}{1.75\rho}\right]^{1/2} \qquad (\text{for } \Re > 1,000) \tag{7-27}$$

In calculating minimum fluidization velocity, the flow regime ($\Re < 1$ or $\Re > 1,000$) should be chosen or guessed; the fluidization velocity should then be determined by using Eq. 7-26 or 7-27 and verified by calculating the Reynolds number after determining the fluidization velocity. If the system is not completely laminar or turbulent, then Eqs. 7-26 and 7-27 cannot be used, and the full set of equations must be solved.

The height of the granular media bed remains constant with the grains supported on one another until the minimum fluidization velocity is reached. As the velocity is increased above the minimum fluidization velocity, the bed height and bed volume increases proportionately as shown in Fig. 7-10.

Example

An 8-m by 3-m sand filter is to be backwashed with water at 20 °C. The sand has a sphericity of 0.95, a diameter of 0.50 mm, and a grain density of 2.1 g/cm³. Assume the porosity is 0.38. What water flow rate is needed to fluidize the media?

Solution

At 20 °C, $\rho = 998.21$ kg/m³ and $\mu = 1.002$ cP $= 1.002 \times 10^{-3}$ kg/m·s.

Start by assuming $\Re > 1,000$, where we can use Eq. 7-27 to calculate the minimum fluidization velocity:

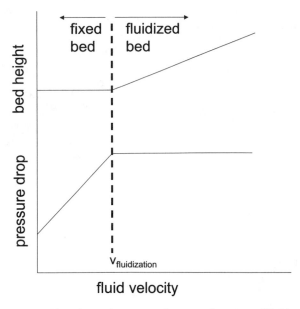

Figure 7-10. Bed height and pressure drop as a function of fluid velocity.

$$V_{\text{fluidization}} = \left[\frac{\phi dg(\rho_g - \rho)\varepsilon^3}{1.75\rho} \right]^{1/2}$$

$$V_{\text{fluidization}} = \left[\frac{0.95 \cdot 5.0 \times 10^{-4}\,\text{m} \cdot 9.81\dfrac{\text{m}}{\text{s}^2} \cdot \left(2100\dfrac{\text{kg}}{\text{m}^3} - 998.21\dfrac{\text{kg}}{\text{m}^3} \right) \cdot (0.38)^3}{1.75 \cdot 998.21\dfrac{\text{kg}}{\text{m}^3}} \right]^{1/2}$$

$$V_{\text{fluidization}} = 0.013\,\frac{\text{m}}{\text{s}}$$

Now we have to check to see whether $\Re > 1{,}000$. From Eq. 7-21, we have

$$\Re = \frac{d \cdot V_{\text{fluidization}} \cdot \tilde{n}}{\grave{\imath}} = \frac{5.0 \times 10^{-4}\,\text{m} \cdot 0.013\dfrac{\text{m}}{\text{s}} \cdot 998.21\dfrac{\text{kg}}{\text{m}^3}}{1.002 \cdot 10^{-3}\dfrac{\text{kg}}{\text{m} \cdot \text{s}}} = 6.3$$

Since the Reynolds number is not greater than 1,000, we have to recalculate using Eq. 7-26.

$$V_{\text{fluidization}} = \frac{g(\rho_g - \rho)}{150\mu} \frac{\varepsilon^3}{1-\varepsilon} \phi^2 d^2$$

$$V_{\text{fludization}} = \frac{9.81\frac{m}{s^2}\left(2100 - 998.21\frac{kg}{m^3}\right)}{150 \cdot 1.002 \times 10^{-3}\frac{kg}{m \cdot s}} \frac{0.38^3}{1-0.38} 0.95^2(5.0 \times 10^{-4}\,m)^2$$

$$V_{\text{fluidization}} = 1.4 \times 10^{-3}\frac{m}{s}$$

Check Reynolds number:

$$\Re = \frac{d \cdot V_{\text{fluidization}} \cdot \tilde{n}}{\grave{\imath}} = \frac{5.0 \times 10^{-4}\,m \cdot 1.4 \times 10^{-3}\frac{m}{s} \cdot 998.21\frac{kg}{m^3}}{1.002 \cdot 10^{-3}\frac{kg}{m \cdot s}} = 0.72$$

Which satisfies the requirement that Reynolds number be less than one. So Eq. 7-26 is a valid equation to use for this case.

The required flow rate for fluidization is

$$Q = 1.4 \times 10^{-3}\frac{m}{s} \cdot (8m \cdot 3m) \cdot \frac{60\,s}{1\,\min} = 2.0\frac{m^3}{\min}$$

A *minimum* flow rate of 2.0 m³/min of water must be supplied for backwashing the filter, and the wash troughs of the filter must be able to handle this flow rate.

Symbol List

a_v	specific surface area
A	wetted surface area of cylinder ($2\pi \cdot$radius$\cdot$length)
d	grain diameter
e	friction factor
f	Kozeny-Carman friction factor
F_K	force on surface by fluid
K	kinetic energy per unit fluid volume ($\frac{1}{2}\cdot\rho\cdot v_{\text{pore}}^2$)
r	cylinder radius
r_H	hydraulic radius
$\Re$	Reynolds number
Q	volumetric fluid flow rate
V	fluid velocity

V_0	superficial velocity
$\mathbf{V}_{bed}$	bed volume
V_{pore}	pore velocity
ε	porosity
ϕ	shape factor, sphericity
ρ	fluid density
ρ_g	grain density
μ	absolute fluid viscosity

Problems

1. Water at 60 °F (15 °C) flows through a 3.5-ft-deep packed bed of 0.016 in. (0.4-mm) diameter granular media with an empty bed velocity of 0.012 ft/s. Assume a porosity of 0.4 and a sphericity of 0.95. What is the pressure loss in psid?
2. Water at 10 °C is flowing through an ion exchange bed. The ion exchange beads are 0.6 mm in diameter with a grain density of 1.15 g/cm^3 and a sphericity of 0.98. The porosity of the bed is 0.38. What is the pressure loss across the bed? What is the minimum fluidization velocity?
3. A granular media filter that is cylindrically shaped (diameter = 2.5 m, depth = 60 cm) is to be backwashed with 18 °C water. The grain diameter is 0.5 mm with a sphericity of 0.85 and a grain density of 3.8 g/cm^3. The porosity of the bed is 0.4. What is the minimum flow rate needed to fluidized the media for backwashing?

References

Droste, R. L. (1997). *Theory and Practice of Water and Wastewater Treatment*, Wiley, New York, NY.

Tchobanoglous, G., and Schroeder, E. D. (1985). *Water Quality*, Addison-Wesley, Reading, MA.

CHAPTER 8

Valves

Chapter Objectives

1. Recommend appropriate valve types for specific applications in treatment systems.
2. Describe the characteristics of the different valve types.
3. Discuss the methods used for valve actuation.
4. Determine the potential for cavitation during fluid flow through valves.

Much effort has been expended over the years to develop valves to control the flow of fluids, as illustrated in Fig. 8-1. The earliest valves were primitive flap valves and plug valves used by the Greeks and Romans to allow or stop water flow. The 1800s and 1900s brought about the invention of more advanced components and designs, including gate, ball, butterfly, and diaphragm valves—designs still in use today. Modern valves are highly developed and are quite sophisticated.

After the discussions about the fundamentals of fluid systems in previous chapters, it should be apparent that in many treatment systems a means to dissipate excess energy is necessary for flow control. This energy dissipation can be done by "throttling" control valves to increase or decrease frictional losses. In addition, in many if not all systems, a means to isolate piping runs by opening or closing a valve is essential. Valves are used for these applications—*flow control valves* and *isolation valves*. The performance of a treatment system depends on the correct application and operation of valves in the system to either control flow rate or isolate piping runs. A significant portion of capital costs associated with constructing a treatment system may be the cost of the valves. Moreover, system operation and maintenance costs can be significantly affected by valve choice. Consequently, proper valve selection is crucial to control costs as well as for proper operation.

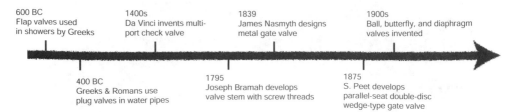

600 BC
Flap valves used
in showers by Greeks

1400s
Da Vinci invents multi-
port check valve

1839
James Nasmyth designs
metal gate valve

1900s
Ball, butterfly, and diaphragm
valves invented

400 BC
Greeks & Romans use
plug valves in water pipes

1795
Joseph Bramah develops
valve stem with screw threads

1875
S. Peet develops
parallel-seat double-disc
wedge-type gate valve

Figure 8-1. Some highlights of valve development. *Source:* Adapted from AWWA (1996).

Valve Categories

The major categories of valves are the following:

Isolation valves are either open or closed and may be used to secure or "isolate" a piping run.

Flow control valves may be throttled to control flow rate or pressure by dissipating energy by friction.

Check valves allow flow in one direction only, and they automatically close to prevent flow in the opposite direction.

Pressure relief valves "relieve" high-pressure conditions in a piping section by opening above a given set point to discharge fluid external to the system.

Pressure control valves are used in a system to separate pipe sections of two different pressures.

Types of Valves

Globe Valves

Globe valves are valves in which a disk is incrementally moved from a seat with valve handwheel rotation, as shown in Fig. 8-2. Globe valves are applicable for flow control as well as isolation. They are suitable for sealing against high pressure (within the working pressure of the valve of course). However, the change in fluid flow direction and magnitude in the valve body and around the disk provides for a significant energy loss, even when fully open. This makes globe valves less desirable than other types (i.e., ball valves) when functioning as isolation valves (i.e., fully opened or closed) when flow control is not needed. It is also common for foreign material to become trapped between the disk and the seat, preventing complete closure when large foreign material is present in the fluid contained by the valve. Solids flowing past the seat can also abrade the seats, although hardened seats are usually used.

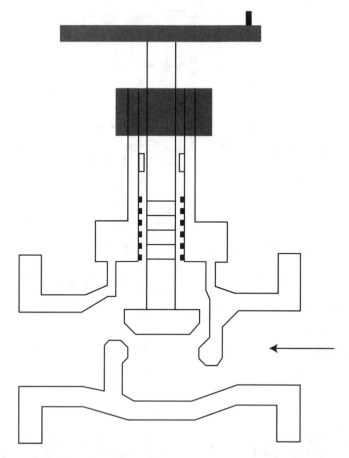

Figure 8-2. Schematic of a globe valve.

Ball Valves

Ball valves consist of a ball within a valve body that can rotate from the closed position to provide an open bore (in the ball) for fluid flow, as shown in Fig. 8-3. Because of the straight open bore in a ball valve when fully open, ball valves furnish low frictional loss when open in comparison to other valves. They can also provide an effective seal against high pressure for complete isolation of the upstream and downstream sides. With a one-quarter turn (90° rotation) of the stem with a handle or actuator, the ball rotates from fully open to fully closed. Ball valves are ideal for isolation valves, but they may also be used for flow control by providing moderate frictional loss when partially open. The ball in smaller ball valves (less than approximately 15 cm or 6 in.) is usually supported by the seats, whereas the larger ball valves usually have trunnions (cylindrical bosses on which the ball pivots) to hold the ball in place (Sanks et al. 1998). A ball valve installed in a water treatment system is shown in Fig. 8-4.

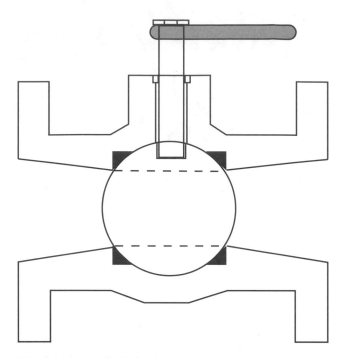

Figure 8-3. Schematic of a ball valve.

Figure 8-4. Ball valve installed in a water treatment system.

Gate Valves

Gate valves are isolation valves that are commonly used in treatment systems. A "gate" or a disk can be fully or partially inserted into the flow stream, as shown in Fig. 8-5, by turning a handwheel, which raises or lowers a stem. Actuators can also be employed. Although some gate valve designs are acceptable for flow control, high velocities can have a deleterious effect on gate valve seats. When fully opened, there is little fluid friction loss as there are only slight perturbations to the flow streams, so gate valves are appropriate for isolation valves. There are double-disk, solid wedge, and resilient-seat gate valves. Double-disk gate valves have two loose disks that press against the seats when closed and are not acceptable for throttling because of their design. Solid wedge gate valves consist of a gate that is

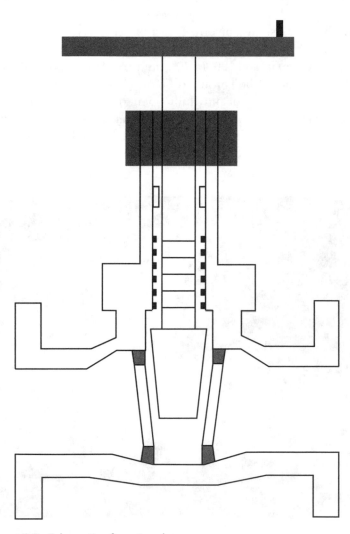

Figure 8-5. Schematic of a gate valve.

much like a solid wedge that fits into the wedge shape of the seats when closed. Resilient-seat gate valves have an elastomer material on the seat or gate for providing a tight seal when closed. Figure 8-6 illustrates a gate valve installed in a water treatment plant.

Butterfly Valves

A butterfly valve is a valve with a disk (within a valve body) that rotates on a shaft as shown in Fig. 8-7. Like a ball valve, butterfly valves can be opened or closed with one-quarter turn of the valve stem (90° rotation). Butterfly valves can be used as isolation valves or as flow control valves in treatment systems. Because the disk in a butterfly valve remains a significant flow restriction even when fully open (see Fig. 8-7), they have a higher energy loss when compared with some other valve types (e.g., ball valves). Because of the disk in the flow stream, the butterfly valve should not be used with high fluid velocities. Butterfly valves are commonly used in water distribution systems, but they do not seal very well at high pressures because of seat compression. A partially open butterfly valve, with the stem opened between approximately 20° to 70°, can provide flow control. However, the valve manufacturer should be consulted before using a butterfly valve for throttling as many are acceptable for only fully opened or fully closed operation. Butterfly valves are relatively simple, being easy to operate and maintain, and are usually lower in cost relative to other valve types. As a result, butterfly valves have

Figure 8-6. Gate valve installed in a water treatment plant.

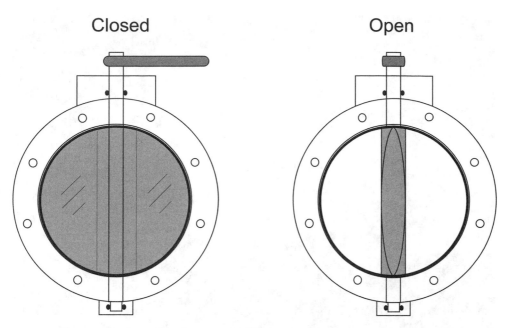

Figure 8-7. Butterfly valve view down axis of pipe/valve.

become very common in treatment systems when appropriate for the application. Figure 8-8 illustrates an installation of a butterfly valve in a water treatment plant.

Plug Valves

Plug valves consist of a plug, shaped like either a cylindrical or a tapered cylinder, that can be turned fully opened or fully closed with a one-quarter turn as for ball and butterfly valves. The plug has a cylindrical bore to allow unimpeded fluid flow, and it can be either full port or reduced port. Plug valves can be used with high pressures because the plug gives a very good seal that provides for isolation across the valve. A plug valve installed in a wastewater treatment plant is shown in Fig. 8-9. A cone valve is similar to a plug valve, but the plug is slightly withdrawn or lifted in the axial direction from the valve bore (body) before rotation (Sanks et al. 1998). After rotation the plug is pushed back into the bore to provide a seal again. This movement of the plug from its seat reduces plug and seat wear, and the torque necessary to cycle the valve.

Check Valves

Check valves allow fluid flow in one direction only. Depending on the design, they have a mechanism for stopping reverse flow. In forward flow, which is in the direction allowed by the check valve, a higher pressure on the upstream side of the valve will generate fluid flow through the valve by opening the check mechanism.

Figure 8-8. Butterfly valve installed in a water treatment system.

Figure 8-9. Plug valves installed in wastewater sludge pump suction lines.

Check valves have significant friction losses with flow, so should be used only when necessary. There are ball check valves, flap valves, swing check valves, lift check valves, and other designs. A schematic of a swing check valve is shown in Fig. 8-10. Figure 8-11 shows an installed swing check valve.

Pressure Relief Valves

Pressure relief valves are used to prevent against excessive pressure in a pipe section, tank, etc. Upon reaching a given threshold pressure, the relief valve will open and allow fluid to pass, alleviating the high-pressure event. A relief valve is shown in Fig. 8-12. The force exerted on the disk must overcome the force countered by the spring for the valve to "crack" open. Usually the spring preload can be adjusted to vary the opening pressure that the valve will relieve.

Recommended Valve Applications

The various valves that are available are not recommended for all services but are appropriate for specific applications. The design of each type of valve renders some valve types inappropriate for some uses. Isolation valves for water service are commonly of ball, butterfly, gate, or plug type. Appropriate isolation valves for wastewater service include ball, gate, or plug valves. Control valves for use in water service are typically of ball, butterfly, globe, or plug type, and control valves for wastewater use include ball or plug valves. Table 8-1 provides a summary of recommended valve applications. Manufacturers' recommendations should always be followed.

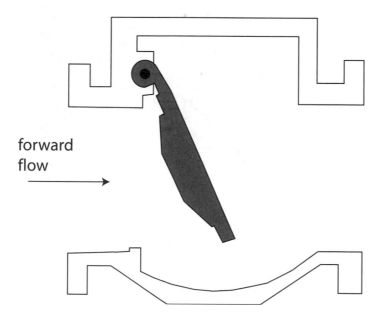

forward
flow

Figure 8-10. Schematic of a swing check valve.

Figure 8-11. Swing check valve installed in a wastewater treatment plant.

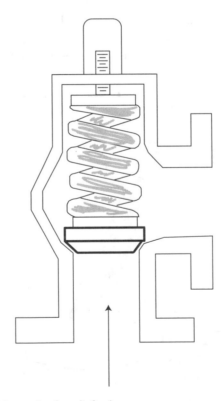

Figure 8-12. Schematic of a relief valve.

Table 8-1. Recommendations for valve applications.

Valve type	Application	Service	
		Untreated water	Untreated wastewater
Ball	Isolation	E	E
	Control	E	E
Butterfly	Isolation	G	N
	Control	F	N
Cone	Isolation	E	E
	Control	E	E
Gate: double disk	Isolation	G	N–F
	Control	N	N
Gate: knife, resilient seat, solid wedge	Isolation	F–G	F–G
	Control	N	N
Globe	Isolation	N	N
	Control	G	N
Plug	Isolation	G–E	G–E
	Control	G–E	G–E

Code: E = excellent, G = good, F = fair, N = not good.
Source: Adapted from Sanks et al. (1998).

When choosing control valves that will be partially open at times, the effect of the percent valve actuation on the flow characteristics of the valve should be considered. As the valve is adjusted from fully closed to fully opened, the friction loss for the flowing fluid and therefore the resulting flow rate may not be linear with respect to valve stroke, as shown in Fig. 8-13. For example, the figure illustrates that a ball valve may not be a good choice to throttle flow between approximately 80% and 100% of maximum actuation (percent of the total one-quarter valve stem rotation) because of the small slope of actuation versus flow rate. Friction loss and flow rate are greatly affected by small changes in valve stroke in this range. A globe valve may be a better choice in this range if accurate flow control is necessary.

As mentioned earlier in the chapter, isolation valves are used to "isolate" components or piping sections. Isolation may be needed on a periodic basis for maintenance of pumps, flowmeters, and other components in the system. Good design practice will include isolation valves at key locations so that disruption to the system is minimized during maintenance operations. Placement of isolation valves should easily allow staff to conduct maintenance work on all components. Isolation may be required at both upstream and downstream connections to a component so that it may be removed or repaired in place. Isolation of certain components may be achieved with only an upstream valve (e.g., fire hydrants). Isolation valves are also necessary to secure redundant or additional piping branches and components (e.g., additional pumps) installed in the system to provide for additional capacity (such as

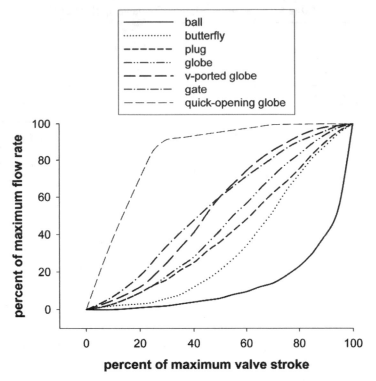

Figure 8-13. Flow rate versus valve stroke for different valve types.
Source: Data from Sanks et al. (1998) and Skousen (2004).

additional flow rate) or to provide for continued operation at design capacity in the event of a component failure, if needed. Often water and wastewater plants operate with entire treatment branches secured until needed. Engineering judgment should be used to discern the location and use of isolation valves in these instances.

Control valves should be correctly sized for the specific application. An undersized control valve produces a larger ΔP than a properly sized valve, thereby consuming greater energy and allowing lower flow rates through the system. If the control valve is undersized by too much, the system may not be able to produce design flow rate. Oversized control valves are an unnecessary cost (since larger valves require more material) and can result in difficult flow regulation as the valve would always be operated in a reduced range of valve stroke—more closed than a properly sized valve. A properly sized control valve is throttled over the full range of stroke available and passes the design flow rate at full open.

Valve Actuators

The different types of valves discussed have different devices to provide actuation to open or close the valves. Actuators for valves can be manual or powered. The requirements for the actuation dictate the type of actuator needed. If the valve needs to be operated remotely, an actuator with remote capability must be chosen. If the valve must be operated without electrical power, electric actuation is not possible unless backup power is supplied. Valve actuators may be of the following types:

- direct manual,
- geared manual,
- electric,
- pneumatic, or
- hydraulic.

Table 8-2 compares the different types of valve actuators. Small one-quarter turn valves (e.g., ball and butterfly valves) can be easily operated with a lever or handwheel attached to the shaft of the valve. When these types of valves are greater than approximately 20 cm (8 in.) in diameter, the operating torques are too high for direct manual operation, and a manual actuator with mechanical advantage (such as gears) is required. Small gate valves (approximately less than 30 cm or 12 in.) can be operated directly through handwheels that contact the valve stem through threads. Again, larger gate valves require torque greater than can be provided directly, and gear reducers are used. The manual actuators for gate valves can be of either rising stem or nonrising stem designs. Pneumatic actuators installed on butterfly valves are shown in Fig. 8-14.

Table 8-2. Valve actuators.

Actuator type	Cost	Characteristics	Requirements	Miscellaneous
Manual	Low	Slow, smooth	Operated at valve	For valves less than approx. 45 cm (18 in.) diameter
Electric	Moderate to expensive	Smooth	Remote, batteries necessary if power failure	
Pneumatic	Moderate	Jerky	Remote, local air receiver	Subject to corrosion from water in air, freezing problems
Hydraulic	Expensive	Very smooth	Remote	Possible corrosion and freezing when aqueous based

Source: Adapted from Sanks et al. (1998).

Figure 8-14. Pneumatic actuators installed on butterfly valves.

Valve Materials

The materials used for seats, bodies, stems, and packing of valves must be carefully selected so as to provide acceptable operation over the longest possible valve lifetime. Obviously, economics must be considered in the choice of materials. The valve body must be strong enough to not deform or rupture under the design pressure of the system in which it is installed. The material must be compatible with the fluid and environment to which the valve is subjected so that excessive corrosion does not shorten its life. Exotic materials may be chosen to provide a very long operating life, but they may not be justified if materials with a lower cost are adequate. Engineering judgment must be used, and the valve manufacturers should be consulted.

The choice of valve body material is dictated by the corrosive conditions expected. Common materials for water and wastewater usage include

- cast iron—ASTM A48, A126; often coated with epoxy (or similar);
- cast steel—ASTM A216; often coated with epoxy (or similar), ~99% iron;
- ductile iron—ASTM A395; often coated with epoxy (or similar); and
- bronze—ASTM B62, B584; alloys of copper, zinc, and other constituents.

The material used for valve seats is extremely important because a corroded or eroded seat will not provide a tight closure. Bronze is very common for valve seats in small valves; however, it is the least able to withstand erosive conditions that may be present in valves, particularly near valve seats. Stainless steel is usually used for valves greater than about 4 in. Stainless steels have great variability in erosion resistance, with 440C being the best alloy for withstanding erosion, and 304 the worst alloy with respect to erosion (Sanks et al. 1998). Stellite seats have excellent service life as the material is very hard and able to resist erosion and corrosion well (Sanks et al. 1998). It is, however, relatively expensive compared to the other seat materials. Resilient seats may provide for a more reliable seal than the rigid seats in the presence of solids in the fluid, but they may degrade over time and in contact with incompatible fluids. Resilient seat materials include the following:

- Buna-N (nitrile) rubber,
- Teflon,
- ultra high molecular weight polyethylene, and
- Viton.

Buna-N (nitrile) rubber is a good, general-purpose seat material that has a relative low cost. Teflon is expensive and very resistant to many chemicals, yet generally has a poor service life. Viton also is resistant to many chemicals but creeps under prolonged use and generally has a poor expected life (Sanks et al. 1998).

The valve stem packing used in valve assemblies is important as it prevents leakage around the valve stem. The packing, inserted around the valve stem, is compressed with a gland via a gland nut. The gland nut can be adjusted to increase force on the gland, which compresses the packing. Usually it is tightened just enough to prevent leakage; if the nut is too tight, high operating torque results. Typical packing materials are

- Teflon fiber,
- acrylic fiber,
- Buna-N (nitrile) rubber, and
- asbestos (which is no longer installed new but may still be present in some installed valves).

Valve Flow Performance

As discussed in Chapter 5, friction from fluid flow in valves is commonly characterized with the flow coefficient, C_V. A valve with a higher C_V has less frictional loss than a valve with a lower C_V as it can pass fluid at a greater flow rate. The values for flow coefficients are provided by the valve manufacturers or can be measured through flow testing.

As discussed in Chapter 5, the pressure loss ΔP through a valve may be calculated from

$$\Delta P = \frac{11.76^2 \cdot Q^2 \cdot SG}{C_V^2} \tag{8-1}$$

for SI units (with flow rate Q in cubic meters per hour and ΔP in kilopascals), where SG is the fluid specific gravity. This equation can be rearranged to

$$Q = \frac{1}{11.76} \cdot C_V \cdot \sqrt{\frac{\Delta P}{SG}} \qquad (8\text{-}2)$$

Plotting Q versus $(\Delta P/SG)^{0.5}$ experimental data yields a straight line, as shown in Fig. 8-15. The slope of the line is equal to $C_V/11.76$ in SI units (Q in cubic meters per hour and ΔP in kilopascals) and the slope is equal to C_V in U.S. Customary units (Q in gallons per minute and ΔP in pounds per square inch).

As the applied pressure is increased to greater values to achieve greater flow through the valve, the relationship between Q and $(\Delta P/SG)^{0.5}$ may deviate from the linear relationship. At greater ΔP, the flow rate through the valve may not be able to be increased above a maximum, Q_{max}. This phenomenon occurs when cavitation takes place. Cavitation is the phase change of a liquid to vapor phase, and it happens when the absolute pressure is below the fluid vapor pressure as discussed in Chapters 2 and 6. It can cause damage to the surfaces in a fluid system as well as cause noise, so it should be avoided. Q_{max} corresponds to "choked cavitation". See Fig. 8-16.

The potential for cavitation to occur with a specific valve in a specific application can be determined by calculating the cavitation index σ, a parameter that represents the ratio of forces resisting cavitation to forces promoting cavitation:

$$\sigma = \frac{P_{downstream} - P_{sat}}{P_{upstream} - P_{downstream}} \qquad (8\text{-}3)$$

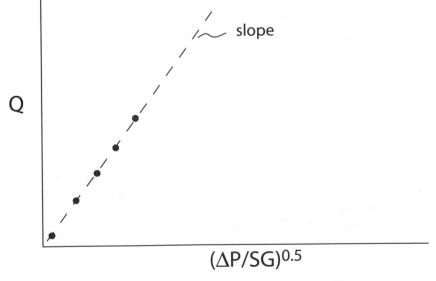

Figure 8-15. Determining C_V from a plot of Q versus $(\Delta P/SG)^{0.5}$.

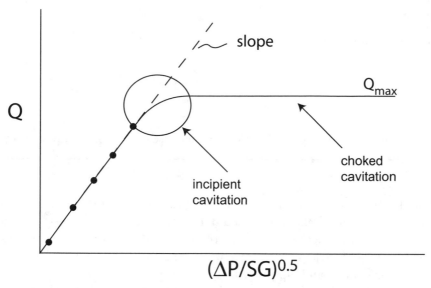

Figure 8-16. Characteristics of incipient and choked cavitation.

A large σ is characteristic of conditions where cavitation does not tend to occur, and a small σ is characteristic of an increased tendency for cavitation. A very large upstream pressure produces a greater opportunity for cavitation, and a small P_{sat} resists cavitation. Typical values for incipient and choked σ are listed in Table 8-3, but valve manufacturers should be consulted for values to be used for specific valves.

Example

Water is passing through a globe valve from below the disk. The upstream pressure is 1,280 kPa absolute pressure (185 psia) and the downstream pressure is 310 kPa (45 psia). The water temperature is 20 °C. Is cavitation expected? The vapor pressure of water at this temperature is 2.3 kPa (0.34 psia).

Table 8-3. Incipient and choked σ for fully open valves.

Valve type	$\sigma_{incipient}$	σ_{choked}
Globe: flow from above disk	0.73	0.38
Globe: flow from below disk	0.52	0.52
Butterfly	3.16	2.19
Ball	5.20	2.19

Source: Data from Skousen (2004).

Solution

The cavitation index is

$$\sigma = \frac{P_{\text{downstream}} - P_{\text{sat}}}{P_{\text{upstream}} - P_{\text{downstream}}} = \frac{310 \text{ kPa} - 2.3 \text{ kPa}}{1,280 \text{ kPa} - 310 \text{ kPa}} = 0.32$$

Therefore cavitation would be expected because $\sigma \ll \sigma_{\text{incipient}}$ and $\sigma \ll \sigma_{\text{choked}}$.

One tactic for reducing the tendency for cavitation is to increase the downstream pressure. Sudden, concentrated, large pressure drops across a valve may result in cavitation. Backpressure control and anticavitation trim in valves can "spread" the friction loss out, preventing large instantaneous friction losses across the valve, and can help prevent cavitation problems. Other approaches to reduce valve cavitation potential are various valve trim or attenuator designs (Skousen 2004), which are valve components that alleviate sharp pressure gradients in the valve.

Symbol List

C_V	flow coefficient
P_{sat}	vapor pressure
Q	flow rate
SG	specific gravity
ΔP	pressure loss across valve
σ	cavitation index

Problems

1. You are the engineer responsible for choosing an isolation valve for a clean drinking water application. What are the appropriate options? Provide brief justification.
2. A control valve is needed for a raw wastewater application. It should be expected that various small solid constituents are present in this wastewater. What valve types are appropriate for this service? Discuss.
3. Water, at 20 °C, is passing through a ball valve with an upstream pressure of 175 psia and a pressure drop across the valve of 40 psid. What is the cavitation index? Is cavitation expected?
4. What changes can be made to the system in Problem 3, specifically to the system pressures, that will increase the cavitation index so that cavitation will be prevented from occurring? Be specific and justify.

Table 8-4. Data for Problem 5.

Flow rate [m³/hr]	Pressure loss [kPa]	Flow rate [m³/hr]	Pressure loss [kPa]
0.9	0.35	4.5	6.9
1.5	0.69	4.8	8.6
1.8	1.7	4.3	10.3
2.7	3.4	4.7	11.7
3.9	4.8	4.5	13.8

5. Flow data for a valve is listed in Table 8-4. Calculate the valve Cv from the data, and determine the choked flow rate.

References

AWWA (1996). *Distribution Valves: Selection, Installation, Field Testing, and Maintenance* (M44), American Water Works Association, Denver, CO.

Sanks, R. L., Tchobanoglous, G., Bosserman, B. E., II, and Jones, G. M., eds. (1998). *Pumping Station Design*, Butterworth-Heinemann, Boston, MA.

Skousen, P. L. (2004). *Valve Handbook*, McGraw-Hill, New York.

CHAPTER 9

Instrumentation

Chapter Objectives

1. Compare and describe the common pressure measuring instruments.
2. Discuss the different types of flowmeters and their characteristics.

The control of a treatment plant requires real-time system data and information to avoid catastrophic failure of system components or failure of the treatment plant to achieve required treatment levels. A control room to monitor and control a large wastewater treatment plant is shown in Fig. 9-1. Instrumentation provides the control room with the information needed to operate the plant. Instrumentation also supplies data at each process location for the operators. Measurement of key parameters helps to ensure proper operation of a treatment system and provides control system input for automatic response and crucial data for troubleshooting system behavior.

The instrumentation installed in a treatment system

- provides continuous measurement of parameters to ensure proper operation,
- supplies input for control systems that automatically respond to certain events,
- can serve to minimize human error, and
- can be useful for troubleshooting problems.

There are many instruments that are used in treatment systems for process control (e.g., ORP, pH, and temperature), but this chapter focuses on instrumentation for monitoring the hydraulics of systems, such as pressure and flow measuring instruments.

Pressure Measurement

Pressure at a point in a fluid system is an important parameter to know for general knowledge of the system operating characteristics, troubleshooting pump

Figure 9-1. Wastewater treatment plant control room.

operation, troubleshooting flow problems, etc. In addition, pressure measurements are frequently used to determine liquid level and flow rate. Common types of pressure measuring instruments are Bourdon tube pressure gauges, diaphragms, and manometers.

Bourdon Tube

The Bourdon tube pressure gauge is the most common pressure indicating instrument (see Fig. 9-2). The Bourdon tube is a C-shaped semicircular length of tube with a flattened oval cross section. When pressure is applied to the inside of the tube (i.e., pressure greater than the pressure on the outside of the tube, which is atmospheric) the tube will straighten. The end of the tube is connected to some device so that the motion, which is a function of the pressure, is converted to a signal. Various devices can be connected to a Bourdon tube, including mechanical indicators, switches, or transmitters. Stainless steels and copper alloys are typical materials used for the Bourdon tube to minimize failure from corrosion. Usually the pressure gauge is filled with glycerin for corrosion protection and to dampen pressure fluctuations.

A recommended Bourdon tube pressure gauge installation is shown in Fig. 9-3. The tap to the system should be in a location where the potential for plugging from solids deposition is minimized, as well as in a location where air will not build up in the tap. For example, the side of a pipe is a good location for a tap, instead of at the top or bottom of the pipe. A snubber (similar to an orifice) is sometimes

Figure 9-2. Bourdon tube pressure gauge.

used in the tubing to the gauge, especially for non-glycerin-filled gauges, to dampen fluctuations. A diaphragm seal may be employed to isolate the fluid in the pressure gauge from the fluid that is being measured. The Bourdon tube pressure gauge is a low-cost pressure measuring apparatus suitable for general use, including in water and wastewater applications.

In lieu of connecting the end of a Bourdon tube to a gauge needle or another type of mechanical indicator, the Bourdon tube may also be coupled to a transducer, as shown in Fig. 9-4. The movement of the Bourdon tube with pressure changes displaces the core element of a linear variable-differential transformer (LVDT). The output voltage of the LVDT can be measured and used to indicate pressure remotely and/or in a control circuit.

Diaphragms

Very thin materials that deflect with minor pressure changes across them may function as a diaphragm pressure measuring device. The force on a diaphragm owing to the differential pressure across it produces a strain that can be directly transmitted to a mechanical movement (although this is somewhat difficult with

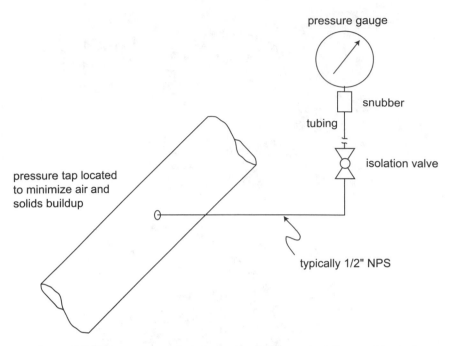

Figure 9-3. Pressure gauge installation in a piping system. *Source*: Adapted from Sanks et al. (1998).

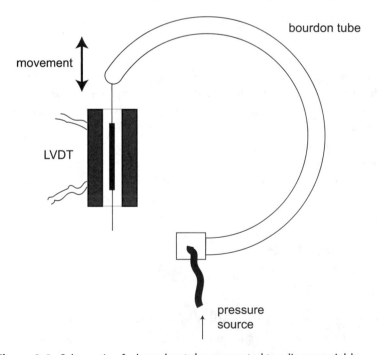

Figure 9-4. Schematic of a bourdon tube connected to a linear variable-differential transformer.

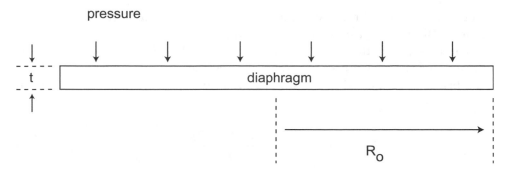

Figure 9-5. Schematic of a thin circular diaphragm used for pressure measurement.

small diaphragm movements) or can be measured with a strain gauge. The materials used for the diaphragm vary widely, and corrosion must be considered as the diaphragm must be very thin for sensitivity. Figure 9-5 shows a schematic of a diaphragm; both tangential and radial strain can result from the pressure differential.

It is typical for the diaphragm to be mounted in a transducer, such as shown in Fig. 9-6, with associated circuitry to provide an output signal (i.e., strain gauges, LVDTs). A strain gauge may be used to determine the amount of strain the diaphragm experiences with pressure across it. The resistance of the strain gauge

Figure 9-6. Pressure transducer.

changes with the diaphragm strain and is measured and translated into pressure readings. Displacement sensors may be used also.

The sensitivity of the signal, which is determined by many factors, is an important issue for choosing a pressure transducer. For an application where a Wheatstone bridge is connected to the strain gauge the sensitivity S can be calculated with (Dally et al. 1984)

$$S = \frac{E_o}{P} = 0.82 \frac{R_0^2(1-\mu^2)E_i}{E \cdot t^2}$$ (9-1)

where E_o is the output voltage, P is the pressure, R_0 is the radius of diaphragm, μ is Poisson's ratio of the diaphragm material, E_i is the driving voltage, and t is the diaphragm thickness.

So a transducer with a greater R_0/t ratio will be more sensitive and provide a greater change in output voltage when used with a Wheatstone bridge for a given pressure change. However, the greater diaphragm diameter results in a larger transducer size.

As an estimate, the resonant frequency of the diaphragm should be three to five times higher than the expected highest frequency of the system pressure change. The resonant frequency of the diaphragm is typically between 10 to 50 Hz and may be calculated with (Dally et al. 1984)

$$f_r = 0.471 \frac{t}{R_0^2} \sqrt{\frac{g \cdot E}{w(1-\mu^2)}}$$ (9-2)

where E is the diaphragm's modulus of elasticity and w is the specific weight of the diaphragm material.

Diaphragms are sometimes connected to a force-balance mechanism where electric power is used to balance the motion of the diaphragm to minimize movement. The output indication for the diaphragm force-balance device is the power necessary to keep the diaphragm stationary. This force-balance mechanism produces little diaphragm movement and wear, prolonging life of the diaphragm.

Manometers

A manometer, comprising a liquid-filled, U-shaped tube with one leg subjected to system pressure and the other leg open to a reference pressure (typically atmospheric), can provide a very simple and effective device for measuring a pressure difference. Manometers may be filled with indicating liquids such as water, sometimes dyed to help visibility, or various oils. Frequently, they are filled with the system fluid, with the leg on the system side completely filled and purged of air, as shown in Fig. 9-7(a), or the indicating liquid may be in contact with a gas above the liquid, as shown in Fig. 9-7(b). With an increased pressure to the system (inlet) of the manometer, the indicating fluid will rise to an increased height in the manometer. The increase in height is related to the difference in pressure exerted

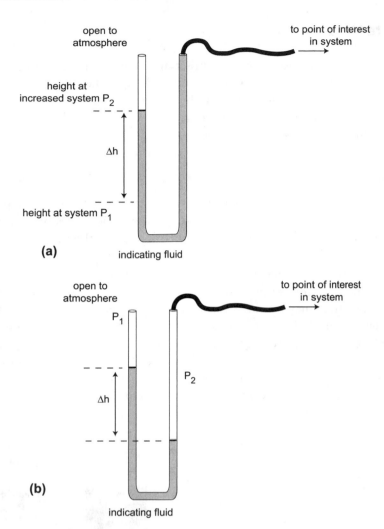

Figure 9-7. Drawings of manometers with (a) the system leg filled with liquid and (b) a gas above the indicating liquid in both legs.

on the manometer inlet. The indicating fluid should be chosen with care; freezing of the liquid may be a concern in cold climates. Using indicating liquids that may release deleterious compounds to the system fluid should be avoided (as in potable water systems). Historically, mercury had been used in manometers in many applications, but its use has been substantially curtailed now because of the health and environmental hazards of mercury.

The pressure increase causing a rise in height (Δh in the indicating leg [the left leg in Figs. 9-7(a) and (b)] may be calculated from

$$\Delta P = P_2 - P_1 = \rho \frac{g}{g_c} \cdot \Delta h \tag{9-3}$$

(see the manometer example in Chapter 3).

Flow Rate Measurement

Flow rate is a crucial measurement in a treatment system. It is important to know the flow rate for properly determining chemical dosages (i.e., chlorine, flocculant, acid, and base) and evaluating adequate hydraulic detention times in the system processes. In this section, the different types of flowmeters that are frequently used in treatment systems are presented.

Orifice Plates

An orifice plate is a simple flow measuring device that may be used in a pipe with pressure gauges to determine flow rate through the orifice. Pressure is measured immediately upstream and downstream of the orifice, and the pressure difference gives an indication of the flow rate. See Fig. 9-8. The measured differential pressure is a function of the velocity squared and can be described with Bernoulli's equation. Solids may get trapped in the orifice or in stagnant areas adjacent to the orifice plate, so they may not be a good choice for liquids with suspended solids (e.g., wastewater or sludge). Orifice plates are very simple and inexpensive, but they have large friction losses. A straight pipe 6 to 45 pipe diameters upstream of the orifice meter is recommended (Sanks et al. 1998) to produce the fully developed flow necessary for accurate and stable pressure readings. A distance of 5 pipe diameters of straight pipe downstream is also recommended.

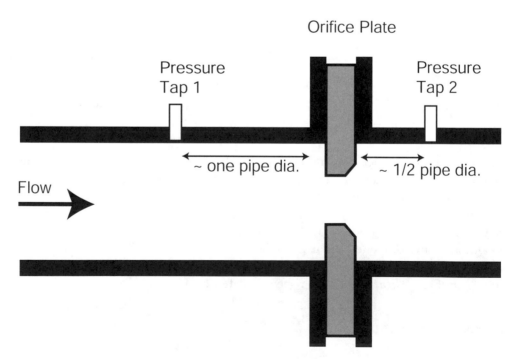

Figure 9-8. Schematic of an orifice plate flowmeter.

Venturi Meter

The venturi meter, comprising a converging section, a throat, and a divergent outlet (see the schematic shown in Fig. 9-9), also produces a differential pressure as a function of the velocity squared. Because of the less abrupt streamline changes, the friction loss of a venturi meter is significantly lower than that for an orifice meter, with approximately 60% less friction loss. Because there is a more "streamlined" flow through the venturi, there is less chance for solid material to become trapped. Because of the converging entrance, venturi meters are a little more forgiving on less than ideal flow profiles in the inlet. Venturi meters, however, are more expensive than orifice meters owing to their greater material cost and increased complexity to manufacture, and require a larger space envelope in the system because of their longer length.

Magnetic Flowmeters

Water passing through a magnetic field produces an electrical potential proportional to the water velocity. Magnetic flowmeters work with electrolytes having conductivities greater than approximately 20 μS/cm (AWWA 1989), such as water. The voltage generated is approximately 0.15 V per m/s ($\sim$0.05 V per ft/s) and is measured as an indication of flow rate. See Fig. 9-10 for a schematic of a magnetic flowmeter and Fig. 9-11 for a magnetic flowmeter installed in a drinking water treatment system. As no protuberances are required that enter into the flow stream to cause a flow restriction, the friction loss is very small when compared with orifice and venturi meters. Certain compounds, such as greases in wastewater, can coat the electrodes and therefore periodic cleaning of the electrodes may be required.

Ultrasonic Flowmeters

There are two types of ultrasonic flowmeters: Doppler and transit-time ultrasonic flowmeters. The Doppler ultrasonic flowmeter measures the frequency *reflected* by solids and gas bubbles traveling in the liquid. The reflected frequency f_r is changed

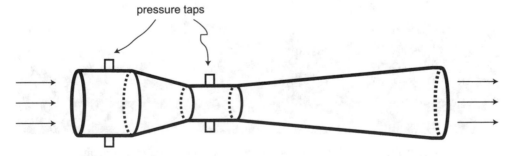

Figure 9-9. Schematic of a venturi flowmeter.

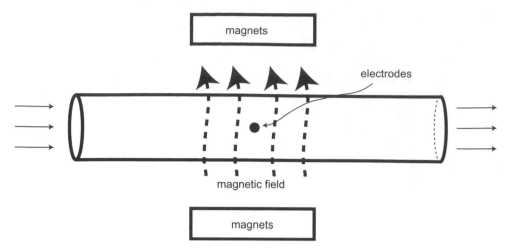

Figure 9-10. Schematic of a magnetic flowmeter.

Figure 9-11. Magnetic flowmeter installed in a drinking water treatment system.

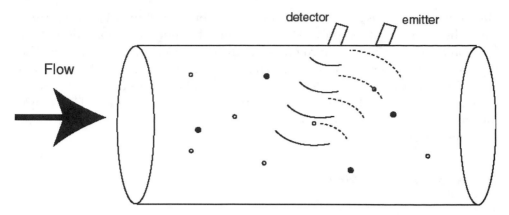

Figure 9-12. Schematic of a Doppler ultrasonic flowmeter.

from the frequency emitted f_e by a piezoelectric transducer by a velocity-dependent function (Sanks et al. 1998)

$$f_r = f_e \left(\frac{v_{sound}}{v_{sound} - v_{solid}} \right) \qquad (9\text{-}4)$$

where v_{sound} is the velocity of sound in the fluid and v_{solid} is the velocity of the solid or bubble.

For the Doppler flowmeter to work a minimum of 100 mg/L of bubbles or solids must be present in the liquid (Sanks et al. 1998). Figure 9-12 shows a schematic of the emitter and detector of a Doppler ultrasonic flowmeter.

A transit-time ultrasonic flowmeter is made up of two transmitters/receivers (transceivers) that are fixed diagonally across a straight pipe section; see Fig. 9-13.

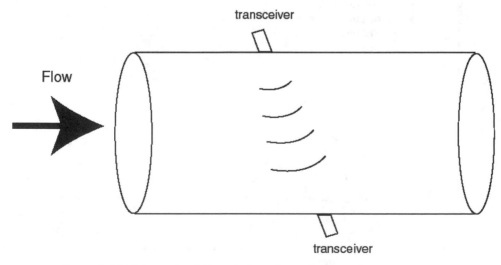

Figure 9-13. Schematic of a transit-time ultrasonic flowmeter.
Source: (Sanks et al. 1998).

One transceiver sends an ultrasonic pulse to the other and the time is recorded. Then the direction of the pulse is reversed, and the time for propagation of the ultrasonic pulse downstream is compared to the pulse propagation upstream. This type of ultrasonic meter is applicable to liquids without an appreciable concentration of impurities. Bubbles, solids, or any other kind of material that will diffract the ultrasonic pulse will interfere with the transit-time measurement. The transit-time ultrasonic meter works well for clean water measurements, but not for wastewater measurements.

Turbine and Propeller Meters

Turbine and propeller meters have blades on a rotor that spins along the centerline of the pipe in which it is installed. The propeller or turbine turns as the fluid passes, and a register or sensor outputs a frequency signal that is an indicator of the flow rate. Because of the rotating element present in the flow stream of the turbine and propeller flowmeters, they have high friction losses. Turbine and propeller meters are only suitable for clean liquids because particulate matter can get into the bearings and cause the meter to fail. As flow rate through the turbine meter increases, the frequency output, voltage output, and friction loss all increase, as shown in Fig. 9-14.

Comparison of Flowmeters

Table 9-1 summarizes some characteristics of the flowmeter types discussed. Flowmeters applicable for treatment systems are available in limited sizes, have a

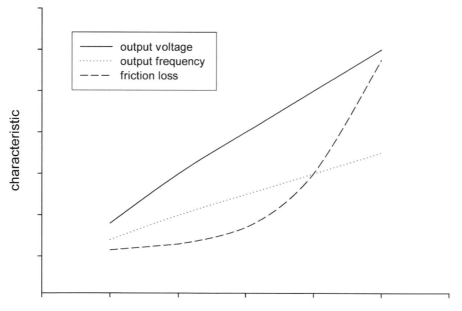

Figure 9-14. Turbine flowmeter signal output and friction loss characteristics.

Table 9-1. Summary of flowmeter types.

Type	Sizes	Head loss	Operating range	Accuracy	Relative cost
Orifice	All	Medium	4:1, 8:1[a]	±0.25% to ±2% of full scale ±1% to ±2%[a]	Low
Venturi	<3 m (10 ft)	Low	4:1 to 10:1	±0.75% ±1% to ±2%[a]	Medium, High[a]
Magnetic	<3 m (10 ft)	None	10:1, 15:1 to 20:1[a]	±0.5% ±0.5% to pm1%[a]	High
Doppler ultrasonic	<3 m (10 ft)	None	10:1[a]	±2% to ± 20%[a]	Low[a]
Transit-time ultrasonic	<3 m (10 ft)	None	20:1	±1% to ±2.5%	High
Propeller	<0.6 m (2 ft)	High	10:1	±0.5% to ±2%	High, Low[a]
Turbine	<0.6 m (2 ft)	High	10:1, 100:1[a]	±0.5% to ±2% ±1%[a]	High

Sources: [a]Data from Sanks et al. (1998). All other data are taken from AWWA (1989).

variety of operating ranges (as low as 4:1 and up to 100:1), are capable of limited accuracies (up to ±20%), and have variable cost. Flowmeters must be chosen carefully, with manufacturer's assistance, to ensure the proper performance of the meter and the system.

For flowmeters to function as designed (and calibrated), the flow entering the flowmeter should be fully developed. To achieve fully developed flow, an adequate length of straight pipe should be provided directly upstream of the flowmeter. Somewhat less critically, there should be a shorter length of straight pipe immediately downstream of the flowmeter. The manufacturer of the flowmeter will specify these lengths and any other critical installation parameter. In the absence of manufacturer's recommendations, a conservative configuration would be to provide 50 pipe diameters upstream of the meter. Flow straighteners may be employed to shorten the required upstream and downstream distances required for proper operation.

The locations of flowmeters in a system should be chosen with care. It is desirable to place flowmeters to collect important flow data, such as in the effluent discharge pipe of wastewater treatment plants or in the pipe discharging from a storage tank. It is not practical to install flowmeters in every piping run, and common, relevant locations should be chosen. As discussed previously, they should also be located as close to where fully developed flow exists in a piping run.

Measuring Flow Rate in Open Channels

Open channels (versus closed conduits or pipes) are frequently used to contain and transport wastewater in treatment systems (see Chapter 12). Weirs and flumes

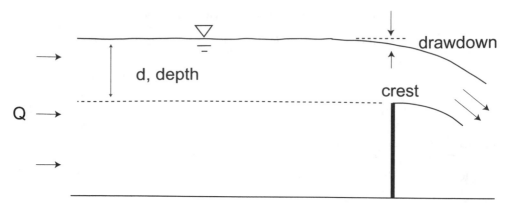

Figure 9-15. Rectangular sharp-crested weir.

are employed in open channel flow to measure flow rate. Weirs consist of a reduction in the channel opening using a geometric-shaped (i.e., rectangular, triangular, etc.) structure perpendicular to the channel and flow axis. The specific geometric shape raises the channel bottom so that the water level is affected by flow, providing for an upstream water depth versus flow rate relationship. In sharp-crested weirs, the water does not contact the downstream face of the weir but separates from the weir at the crest.

A rectangular sharp-crested weir is illustrated in Fig. 9-15. For a rectangular channel cross section, the flow rate may be calculated with

$$Q = C_w \cdot T \cdot d^{3/2} \tag{9-5}$$

where C_w is the weir coefficient, T is the channel (and weir) width, and d is the water depth above the crest. A typical C_w for a sharp-crested rectangular weir is approximately 1.84 $m^{1/2}$/s (3.33 $ft^{1/2}$/s) (U.S. Department of the Interior 1997).

For a triangular sharp-crested or V-notch weir with the cross section shown in Fig. 9-16, the flow rate may be calculated from

$$Q = C_w d^{5/2} \tag{9-6}$$

A typical C_w for a V-notch weir is 1.38 $m^{1/2}$/s (2.49 $ft^{1/2}$/s) (U.S. Department of the Interior 1997).

Broad-crested weirs are also used for flow rate measurement in open channels. The water flow over a broad-crested weir does not separate from the downstream edge of the weir as it does in flow over a sharp-crested weir. A rectangular broad-crested weir is shown in Fig. 9-17. The flow rate versus depth relationship is described with Eq. 9-5.

Flumes consist of a reduction in channel width and a change in the channel bottom slope that brings about a relationship between flow rate and water depth

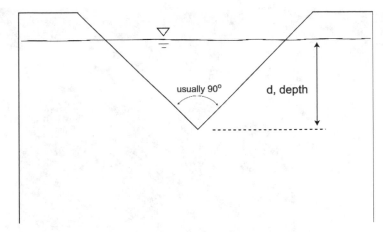

Figure 9-16. Cross section of triangular sharp-crested weir.

at a given point in the flume. The depth of water at specific locations immediately upstream of the weir or flume throat is monitored with staff gauges, floats, or bubble gauges, or ultrasonically (as in Fig. 9-18). The water flow rate through the weir or flume can then be determined from that water depth using known correlations, some of which have already been noted (i.e., Eqs. 9-5 and 9-6). Parshall flumes, configured so that supercritical flow is produced in the throat, are used in many applications for determining flow rate in open channels. (See Chapter 12 for a discussion on supercritical flow.) Because of the configuration of the Parshall flume, only one water depth measurement is needed to ascertain the flow rate through the flume. Published flow rate versus water depth relationships exist for flumes (see Kilpatrick and Schneider 1983; U.S. Department of the Interior 1997). A Parshall flume installed at a wastewater treatment facility is shown in Fig. 9-18.

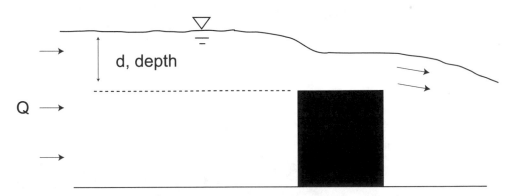

Figure 9-17. Rectangular broad-crested weir.

Figure 9-18. Parshall flume in a wastewater treatment system.

Symbol List

C_w	weir coefficient
d	water depth above crest
E	modulus of elasticity
E_i	driving voltage
E_o	output voltage
f_e	frequency produced by ultrasonic flowmeter emitter
f_r	resonant frequency, reflected frequency
P	pressure
R_0	radius of diaphragm
t	diaphragm thickness
T	channel or weir width
v_{sound}	velocity of sound in fluid
v_{solid}	velocity of solid or bubble
w	specific weight of the diaphragm material
μ	Poisson's ratio

Problems

1. A point in a treatment system is expected to have an internal pressure ranging from atmospheric up to 1.4 kPa above atmospheric pressure (0.20 psig). You are planning on measuring the pressure with a manometer and are considering various indicating fluids. What are the Δh values in the manometer for water (0.998 g/cm^3), ethylene glycol (1.11 g/cm^3), silicone oil (0.961 g/cm^3), and mercury (13.5 g/cm^3)? What would the Δh values be for an increase in pressure to 20 kPa above atmospheric (2.9 psig)?

2. Determine the ratios of output voltage to the driving voltage at 75 psi for 0.015-in.-thick stainless steel diaphragms that have diameters of (a) 3/8 in. and (b) 1½ in. How does the output voltage to the driving voltage ratio compare for the two diaphragm diameters? Take μ as 0.25 and E to be 28.5×10^6 psi.

3. You are the engineer designing a wastewater treatment system, and you need a flowmeter in a 0.5-m-diameter raw sewage line. What are your options? Choose a flowmeter and provide justification.

4. A weir is to be installed in a channel to provide a method for measuring flow rate. A rectangular sharp-crested weir and a triangular (90°) sharp-crested weir are being considered. The channel is 4 m wide and the maximum flow rate is 7 m^3/s. Because of site conditions, the height of the water above the crest cannot exceed 1 m. Will a rectangular sharp-crested weir or a triangular (90°) sharp-crested weir (or both) suffice in this situation without exceeding the crest height? Justify your answer.

References

AWWA (1989). *Flowmeters in Water Supply, M33*, American Water Works Association, Denver, CO.

Dally, J. W., Riley, W. F., and McConnell, K. G. (1984). *Instrumentation for Engineering Measurements*, Wiley, Hoboken, NJ.

Kilpatrick, F. A., and Schneider, V. R. (1983). *Techniques of Water-Resources Investigations of the United States Geological Survey*, Chapter A14, Use of Flumes in Measuring Discharge, U.S. Geological Survey, Alexandria, VA.

Sanks, R. L., Tchobanoglous, G., Bosserman, B. E., II, and Jones, G. M., eds. (1998). *Pumping Station Design*, Butterworth-Heinemann, Boston, MA.

US Department of the Interior, Bureau of Reclamation (1997). Water Measurement Manual, US Government Printing Office, Washington, DC.

CHAPTER 10

Piping Materials and Corrosion

Chapter Objectives

1. Identify common piping material for treatment plants.
2. Summarize fundamentals of corrosion in fluid systems.

Piping Material

Many different fluids must be conveyed from place to place in treatment systems. These fluids may be corrosive, hazardous, or flammable or may contain pathogens. Different materials have been used for piping materials in treatment systems; these include ferrous, copper, and cement-based materials as well as thermoplastics. Piping systems must operate for long periods of time and are expected to require minimal repair. The inside of a cast iron water main that has successfully operated for over 100 years, illustrated in Fig. 10-1, shows a buildup of deposits that may increase the frictional loss of the main. The water main is in remarkably good condition for 100 years of use.

Ferrous Materials

Ferrous materials include cast iron, ductile iron, and steel, both zinc-coated (galvanized) and stainless. Often the iron and steel piping have liners of materials that are more resistant to corrosion and erosion than the base ferrous material. Lining material includes cement for sewage and activated sludge, glass for exceptionally corrosive fluids, cement or epoxy for potable water, and polyurethane and polyethylene for sewage. Lined steel or ductile iron pipe is typical for large sewage piping. Stainless steels have iron, carbon, chromium, and sometimes other constituents (e.g., Mo or Ni).

Copper Alloys

The copper alloys used for piping in treatment systems are usually brasses (copper and zinc alloys), gunmetals (copper, tin, and zinc alloys), copper–nickel alloy (70:30 copper:nickel), and Monel (70:30 nickel:copper alloy).

Figure 10-1. Interior of a 48-in. cast-iron water main more than 100 years old. *Source:* The author gratefully acknowledges Camp Dresser & McKee, Inc., and the City of New Bedford, Massachusetts, for permission to use photographs of the City of New Bedford's water distribution system. © Camp Dresser & McKee Inc. All rights reserved. Used by permission.

Cement-based Materials

Cement-based materials are also used for piping; these comprise aggregate (sand or sand and gravel) and a binder. The binders are commonly Portland cement or high alumina cement.

Thermoplastic Materials

Polyvinyl chloride (PVC) is the most common plastic pipe material used in the United States. Other materials include polyethylene (PE), high-density polyethylene (HDPE), acrylonitrile-butadiene-styrene (ABS), chlorinated polyvinyl chloride (CPVC), polypropylene (PP), polybutylene (PB), and fiberglass-reinforced plastic (FRP). Thermoplastic materials are used for water service, sewage, sludge, and other corrosive fluids. However, some materials are not suitable for all installations. Ultraviolet light can cause deterioration of some plastics over time, and some thermoplastics can be attacked by various chemicals (e.g., gasoline constituents) or may at least be permeable to some compounds (which is important if a plastic potable water pipe runs through ground that is contaminated with a gasoline or solvent spill). Plastic pipe should never be used for compressed gases.

Corrosion

Corrosion Principles

Corrosion is the physical and chemical effect of the environment on a material. Corrosion of metal pipe and piping components results in leakage, infiltration, higher friction losses from increased roughness and from deposition of products of corrosion, and degradation of water quality (e.g., dissolution of lead, copper, and other corrosion byproducts into drinking water). Corrosion occurs through various mechanisms, such as crevice corrosion and galvanic corrosion (which will be discussed in the following section), but all mechanisms are based on fundamental oxidation–reduction or *redox* chemistry.

Redox reactions are electron transfer reactions. An oxidation half-reaction is one in which a substance loses electrons, and a reduction half-reaction is one in which a substance gains electrons. These half-reactions must happen simultaneously since electrons cannot exist freely in solution. There must be at least one oxidation reaction and at least one reduction reaction to form a redox "couple." One species is reduced while one is oxidized.

General oxidation–reduction half-reactions may be shown as

$$\text{reduction:} \quad \text{ox} + ne^- \rightarrow \text{red} \quad\quad (10\text{-}1)$$

$$\text{oxidation:} \quad \text{red} \rightarrow \text{ox} + ne^- \quad\quad (10\text{-}2)$$

where ox is the oxidized species, red is the reduced species, e^- is an electron, and n is the number of electrons transferred.

For this redox couple to take place, an electrochemical cell must be present. The electrochemical cell must have the following:

- an anode, where the oxidation reaction(s) takes place,
- a cathode, where the reduction reaction(s) takes place,
- a path for electron transport, and
- an electrolyte solution for conducting ions.

If any of these components is not present, the oxidation and reduction reactions cannot take place. Each of these necessary components is shown in Fig. 10-2 for an idealized electrochemical cell.

For corrosion of a metal (Me), the metal is oxidized to a lower oxidation state:

$$\text{Me} \leftrightarrow \text{Me}^{n+} + ne^- \quad\quad (10\text{-}3)$$

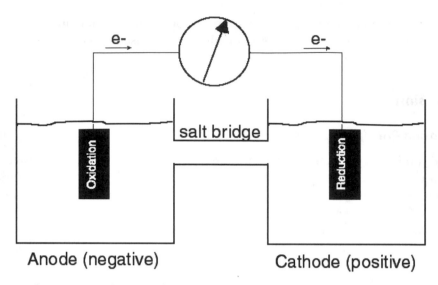

Figure 10-2. Schematic of an electrochemical cell.

The oxidized metal may be more soluble than the original, reduced metal, thereby depleting the amount of metal in the pipe and resulting in more metal dissolving into the water. It may have different properties that may hinder further corrosion or may have undesirable properties. An example of iron oxidation reaction that depletes the base metal is

$$Fe^0(s) \rightarrow Fe^{2+} + 2e^- \tag{10-4}$$

The ferrous iron in the ionic form is soluble and mobile in the water phase. So iron in contact with water would be depleted over time when this oxidation reactor occurs. The ferrous iron may be further oxidized to ferric iron:

$$Fe^{2+} \rightarrow Fe^{3+} + e^- \tag{10-5}$$

Dissolved ionic ferrous and ferric iron may undergo complexation reactions, producing oxides, hydroxides, carbonates, and other complexes (with other ligands), potentially forming corrosion scale as discussed in the following.

An electrochemical cell where metal corrosion occurs in a system is termed a *corrosion cell*. A schematic of an idealized corrosion cell is in Fig. 10-3, which shows the "base" metal corroding at the anode and possible reduction reactions taking place at the cathode.

For a corrosion reaction of metals in contact with water, the extent of the reaction, which defines whether or not corrosion takes place, depends on the aqueous conditions and on the metal in contact with the water. The potential for corrosive oxidation reactions can be quantified with E_H, the redox potential. A high E_H corresponds to highly oxidizing conditions (e.g., high dissolved oxygen concentration), and a low E_H corresponds to highly reducing conditions. A plot of the stable

metal oxidation reaction:

various reduction reactions:

$$2H^+ + 2e^- \rightarrow H_2$$
$$O_2 + 2H_2O + 4e^- \rightarrow 4OH^-$$
$$HOCl + 2e^- \rightarrow Cl^- + OH^-$$

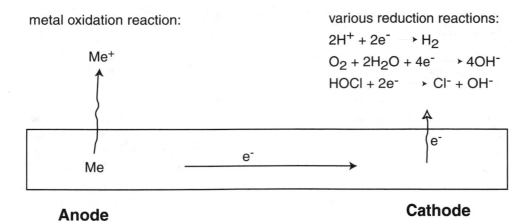

Anode   **Cathode**

Figure 10-3. Idealized corrosion cell. Me is a corroding metal at the anode.

metal species as a function of E_H and pH is called a Pourbaix diagram. A Pourbaix diagram for iron is shown in Fig. 10-4. For highly reducing conditions, Fe^0, the zero-valent solid phase iron, is the stable form of iron present. However, as conditions become more oxidizing, iron is present in higher oxidation states. For example, at lower pH, Fe^{2+} and Fe^{3+}, both dissolved species, are the dominant

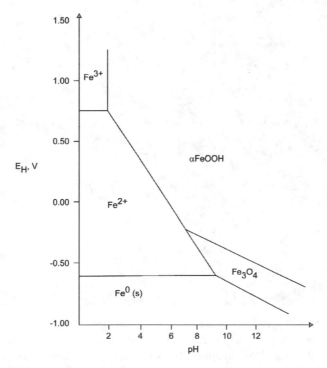

Figure 10-4. Iron Pourbaix diagram showing predominant species as a function of E_H and pH.

forms of iron. Therefore at conditions promoting the dissolved species, removal of iron takes place and iron *corrosion* occurs.

Corrosion Scales

Under some conditions as illustrated in Fig. 10-4, the solids αFeOOH (goethite) and Fe_3O_4 (magnetite) may be formed from other forms of iron (Fe^0, Fe^{2+}, or Fe^{3+}). A steel pipe corroding at conditions promoting these (or other) solids may have a layer of these solids build up on the surface as a scale. See Fig. 10-5. Other solids such as $FeCO_3(s)$ (siderite), $Fe(OH)_2(s)$, and hydrated ferric oxides can form a scale on steel and iron surfaces as well. These solids are the result of products of corrosion of the base metal. Figure 10-6 shows the various materials that may form in scale on a cast iron pipe.

With time, as corrosion of the base metal proceeds, the scale layer increases in thickness and can effectively retard the rate of corrosion. With the development of scale layers such as this the corrosion rate can decrease significantly with time. See Fig. 10-6.

Figure 10-5. Layers of scale on a section of iron pipe from a 16-in. water main. *Source:* The author gratefully acknowledges Camp Dresser & McKee, Inc., and the City of New Bedford, Massachusetts, for permission to use photographs of the City of New Bedford's water distribution system.

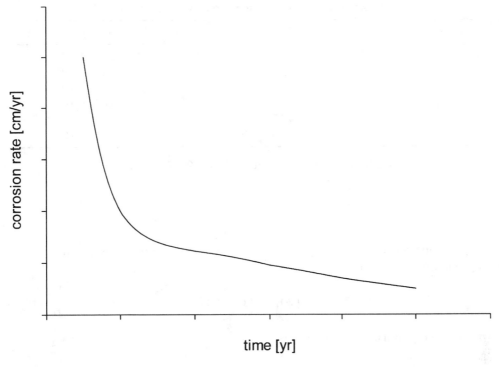

Figure 10-6. Graph of corrosion rate versus time.

The formation of a corrosion-retarding layer is also called *passivation*. The reduction in corrosion rate is thought to be because of the reduced mass transport rate of oxygen through the scale layer to the base metal, which thereby reduces the cathodic reaction rate. As the reactions at the anode and cathode must operate at the same rate (because electrons cannot exist freely in solution), the overall corrosion rate is therefore reduced by this scale.

Calcium carbonate scale can also protect a pipe from corrosion. The scale is formed through the reaction

$$Ca^{+2} + CO_3^{-2} \underset{\leftarrow}{\overset{\rightarrow}{}} CaCO_3 \qquad (10\text{-}6)$$

The equilibrium conditions for this reaction can be predicted with the solubility product for calcium carbonate formation:

$$K_{sp} = [Ca^{+2}][CO_3^{-2}] \qquad (10\text{-}7)$$

where the brackets, [ ], designate the concentration of the species in moles per liter. The solubility product for calcium carbonate is a function of temperature and values are listed in Table 10-1.

Table 10-1. Carbonate system equilibrium constants.

Temperature [°C]	K_{sp} [mol^2/L^2]	K_{carb} [mol/L]
5	8.13×10^{-9}	2.75×10^{-11}
10	7.08×10^{-9}	3.24×10^{-11}
15	6.03×10^{-9}	3.72×10^{-11}
20	5.25×10^{-9}	4.17×10^{-11}
25	4.57×10^{-9}	4.68×10^{-11}
40	3.09×10^{-9}	6.03×10^{-11}

Source: Data are from Tchobanoglous and Schroeder (1985).

Carbonate–bicarbonate equilibrium must be accounted for also, as the following reaction will occur:

$$HCO_3^- \underset{\leftarrow}{\overset{\rightarrow}{}} H^+ + CO_3^{-2} \tag{10-8}$$

where the equilibrium expression is

$$K_{carb} = \frac{[H^+][CO_3^{-2}]}{[HCO_3^-]} \tag{10-9}$$

For a given concentration of CO_3^{-2}, raising the pH of the water will cause calcium carbonate precipitate to form and coat the piping material with a protective coating. The Langelier Index (LI) is a well-accepted indication for predicting the scaling behavior of calcium carbonate. It is calculated by

$$LI = pH_{measured} - pH_{sat} \tag{10-10}$$

where $pH_{measured}$ is the actual, measured pH of the water and pH_{sat} is the pH of the water in equilibrium with solid-phase $CaCO_3$.

The value for pH_{sat} is calculated with (Tchobanoglous and Schroeder, 1985):

$$pH_{sat} = -\log\left(\frac{K_{carb} \cdot \gamma_{Ca^{+2}}[Ca^{+2}] \cdot \gamma_{HCO_3^-}[HCO_3^-]}{K_{sp}} \right) \tag{10-11}$$

where $\gamma_{Ca^{+2}}$ is the activity coefficient for Ca^{+2} and $\gamma_{HCO_3^-}$ is the activity coefficient for HCO_3^-. Activity coefficients are used to account for nonideality of the ions in solution (since dissolved ions are affected somewhat by other ions in solution and are not just surrounded by water molecules). As a first approximation, the activity coefficients can be assumed to be 1.0, but for a more accurate calculation of the

Table 10-2. The Langelier Index.

Langelier Index (LI)	Water characteristic	$CaCO_3$ saturation
LI > 0	Scale forming	Supersaturated
LI = 0	Neutral	
LI < 0	Corrosive	Undersaturated

activity coefficients, use the following equation to calculate the activity coefficient for each species:

$$\log \gamma_i = -\frac{0.5 \cdot (z_i)^2 \cdot I^{1/2}}{1 + I^{1/2}} \tag{10-12}$$

where z_i is the charge of species i and I is the solution ionic strength.

The calculated value for the Langelier Index is compared to values listed in Table 10-2 to determine whether the water is scale forming (encrustive) or corrosive. A positive LI indicates scale has a tendency to form and protect the pipe and fittings from corrosion, although excessive scaling can be problematic if it occurs.

Forms of Corrosion

Uniform Corrosion

For a metal surface of spatially identical physical and chemical characteristics, the surface corrodes at a uniform rate. Local sites on the surface function as both anodes and cathodes at different times. The sites that function as anodes can instantaneously change to become cathodes, and vice versa. So a local site can function as an anode one moment and as a cathode the next moment. The corrosion, or dissolution of the metal, takes place at the anode. As the corrosion sites move around on the surface of the metal, the metal depletion is relatively uniform. In actuality, the corrosion rate does not actually turn out to be completely uniform. Imperfections in the chemical composition and/or the crystal structure of the metal can provide local sites that may preferentially function as anodes or cathodes. Variations in scale layers can also provide for a greater tendency for a local site to have either oxidation or reduction reactions.

Galvanic Corrosion

If an electrochemical cell consists of two electrochemically different metals, one metal functions as an anode and the other as a cathode, resulting in *galvanic corrosion*. Galvanic corrosion is the result of the use of two galvanically dissimilar metals in a system, one more noble than the other. The less noble material serves as an anode and is preferentially corroded, while the more noble metal functions as

Table 10-3. The galvanic series.

Anodic, least noble	Magnesium, magnesium alloys
	Zinc
	Aluminum alloys
	Cadmium
	Mild steel, cast iron
	Iron alloys
	Lead, tin, lead–tin solders
	Nickel
	Brass, copper, bronze
	Titanium, monel (Ni–Cu alloy), silver solder
	Silver, gold
Cathodic, most noble	Platinum, graphite

the cathode. The reduction reactions occur at the more noble metal surface (cathode), and the oxidation reactions occur at the less noble surface (anode). It is important for a fluid system designer to be aware of the coupling of galvanically dissimilar metals in treatment systems. The galvanic series listing the relative nobility of various metals is given in Table 10-3.

Coupling metals from different levels in the galvanic series results in preferential corrosion of the least noble metal. The further apart on the galvanic series the metals are, the greater is the potential for corrosion of the less noble occurs. Relative surface area also plays a role in galvanic corrosion. A small surface area for the less noble metal relative to the surface area of the more noble metal results in greater corrosion of the less noble material.

A zinc coating on steel produces *galvanized* steel. Because zinc is less noble than steel, it is preferentially corroded with respect to the steel and any uncoated steel areas will have a reduced corrosion rate. The steel is thus protected, not just by being coated, but also by the galvanic corrosion of the zinc. Figure 10-7 shows

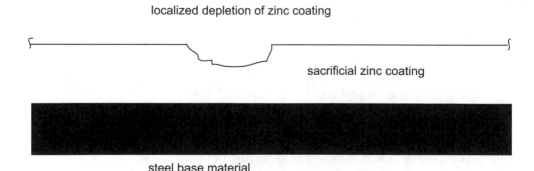

Figure 10-7. Drawing showing the corrosion of a zinc coating on steel.

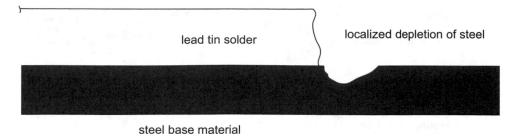

lead tin solder

localized depletion of steel

steel base material

Figure 10-8. Drawing showing sacrificial galvanic corrosion of a less noble steel base material.

the effect of corrosion when steel is coated with zinc, a less noble material than the steel. Figure 10-8 shows the effect of a galvanic couple composed of steel and lead–tin solder. The steel is preferentially depleted by the galvanic couple with the more noble lead–tin solder.

Localized Corrosion

Localized areas on the surface of metal experiencing corrosive conditions commonly have some imperfections in the base metal and/or discontinuities in passivating scales or coatings. Surface areas with imperfections usually function as anodes, with oxidation reactions taking place there, producing localized corrosion.

Metal exposed to continuous stresses are also subject to enhanced corrosion because these local areas are anodes in a corrosion cell. This is termed *stress corrosion*. In this case the anodic surface area is usually much smaller than the surface area of the cathode. The corrosion in these small anodic stress areas can produce localized corrosion at a rate much greater than experienced by the surrounding areas.

Concentration Cell Corrosion

As discussed earlier, for corrosion reactions to proceed at an anode in a corrosion cell, reduction reactions must also simultaneously occur at a cathode. The potential for these reduction reactions is a function of the presence and concentrations of aqueous species such as oxygen and/or hydrogen ions. Different concentrations of these species at local sites can affect the overall corrosion reactions and produce a localized corrosion at a much higher rate than uniform corrosion. This is called *concentration cell corrosion*. It is very common for different concentrations of dissolved oxygen (DO) to form a *concentration cell*; for different oxygen concentrations it is more specifically called *differential oxygenation corrosion*. The metal area adjacent to the higher DO concentration functions as the cathode, whereas the area adjacent to the low DO functions as the anode. Significant corrosion occurs

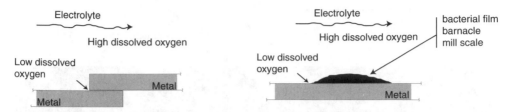

Figure 10-9. Physical configurations that may produce differential oxygenation corrosion. *Source:* Adapted from AWWA (1996).

at the anode, which is the low DO area. Any physical configuration that can reduce or prohibit the mass transport of oxygen into an area can result in differential oxygenation corrosion. This type of corrosion can be caused by rivets, bolts, mill scale, bacterial slime, debris, gasketed joints, slip joints, pipe threads, socket-type joints, etc. Tubercles (deposits of corrosion products) may provide for a local environment depleted in oxygen that can produce differential oxygenation corrosion also. Examples of physical configurations that may produce this accelerated corrosion are shown in Fig. 10-9.

Reducing Corrosion

The best technique to control corrosion is to use materials that provide for an acceptable service life for the anticipated conditions. Proper material selection is conducted through knowledge of corrosion principles, familiarity with the available materials and coatings, and engineering experience and judgment. Certain system conditions are desirable to avoid excessive corrosion in piping systems. See Table 10-4.

Table 10-4. Water chemistry for minimizing corrosion in specific materials.

Material	Corrosion type	pH	Alkalinity [meq/L]	Other
Cement	Uniform	>7	0.3 to 0.5	Calcium > 10 mg/L
Iron, steel	Uniform	>7	0.2 to 0.5	
	Pitting			Dissolved oxygen > 2 mg/L
Copper	Uniform and cold-water pitting	>7		
	Hot-water pitting	>7	1 to 2	

Source: Adapted from AWWA (1996).

Corrosion in water systems may be controlled somewhat by modifying the water chemistry to within the ranges listed in Table 10-4 for specific materials. Changes can be made to solution pH, alkalinity, and other properties. Chemicals typically used for adjusting pH and alkalinity include lime [$Ca(OH)_2$], caustic soda ($NaOH$), soda ash (Na_2CO_3), sodium bicarbonate ($NaHCO_3$), and carbon dioxide. The addition of 1 mg/L of lime adds 1.35 mg $CaCO_3$/L alkalinity, 1 mg/L of 50% caustic soda adds 1.25 mg $CaCO_3$/L alkalinity, 1 mg/L of soda ash adds 0.94 mg $CaCO_3$/L alkalinity and 1 mg/L of sodium bicarbonate adds 0.59 mg $CaCO_3$/L alkalinity (AWWA 1996). The chemicals can also be used to adjust the Langelier Index to form scale (encrustive conditions). Chemicals that function as corrosion inhibitors can also be added to the system. Common corrosion inhibitors that are added to water distribution systems are shown in Table 10-5.

For control of external corrosion, a *cathodic protection sacrificial anode system* such as that shown in Fig. 10-10 can be used. In the sacrificial anode system, a material less noble than the cathode (the protected metal) is used as the anode. Zinc and magnesium are common anode materials for use in cathodic protection. Because they are less noble than the pipe that is being protected, they preferentially corrode. Corrosion does not occur at the pipe because it is functioning as a cathode—reduction reactions take place at the cathode. Note that the four components necessary for a corrosion cell are required here as well: the anode, the cathode, a path for electrons, and an electrolyte (water).

An electrochemical cell (corrosion cell) to protect piping can be formed by passing a current from the cathode to anode, as illustrated in Fig. 10-11. This is called an *impressed-current cathodic protection system*. The electrical current actually forms the anode and cathode; a coupling of more noble and less noble metals is not needed. Corrosion occurs at the anode, as for all corrosion reactions, and the cathode is protected through reduction reactions.

Table 10-5. Corrosion inhibitors for drinking water systems.

Chemical	Forms	Typical strength
Sodium silicate	Viscous liquid	38% to 42%
Zinc orthophosphate	Liquid	
Sodium hexametaphosphate	Solid plates, flakes, lumps, granular	>80% PO_4
Sodium tripolyphosphate	Powder, granular	>75% PO_4
Monobasic sodium phosphates	Crystalline powder or granular	>77.7% PO_4
Dibasic sodium phosphates	Crystalline powder or granular	>64.3% PO_4

Source: Adapted from AWWA (1996).

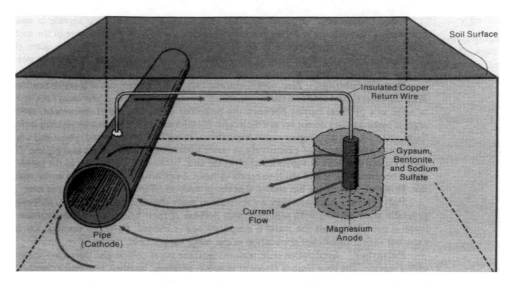

Figure 10-10. Drawing of a cathodic protection system with sacrificial anode. *Source:* AWWA (2004), with permission from American Water Works Association.

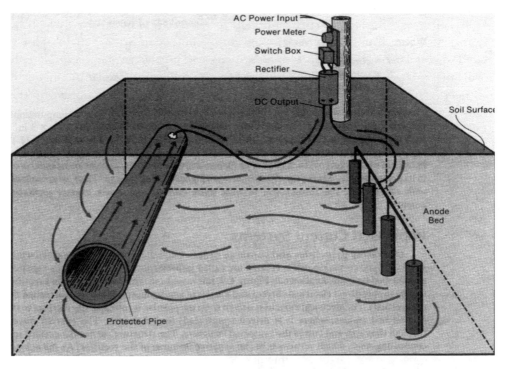

Figure 10-11. Drawing of an impressed-current cathodic protection system. *Source:* AWWA (2004), with permission from American Water Works Association.

Symbol List

E_{H}	redox potential
I	solution ionic strength
K_{carb}	solubility product for carbonate–bicarbonate equilibrium
K_{sp}	solubility product for calcium carbonate formation
LI	Langelier Index
z_i	charge of species i
$\gamma_{\mathrm{Ca}^{+2}}$	activity coefficient for Ca^{+2}
$\gamma_{\mathrm{HCO_3^-}}$	activity coefficient for HCO_3^-
γ_i	activity coefficient for species i

Problems

1. Water at 5 °C has a pH of 7.0, $[Ca^{2+}] = 2.1$ mM, and $[HCO_3^-] = 2.2$ mM. What is the Langelier Index? Will a protective scale be formed? What can be done to the chemistry to increase the LI such that scale would be expected?
2. Would it be considered good practice from a corrosion standpoint to install a threaded steel plug in a threaded hole in Ni–Cu alloy pipe? Why? What about installing a Ni–Cu threaded plug in a steel pipe?
3. Pitting is observed in copper pipe carrying pH = 6.2 water. What do you suggest as a solution?

References

AWWA (1996). *Internal Corrosion of Water Distribution Systems*, American Water Works Association, Denver, CO.

AWWA (2004). *External Corrosion: Introduction to Chemistry and Control, M27*, 2nd Ed., American Water Works Association, Denver, CO.

Tchobanoglous, G., and Schroeder, E. D. (1985). *Water Quality*, Addison-Wesley, Reading, MA.

CHAPTER 11

Fluid Flow Transients

Chapter Objectives

1. Assess the causes of hydraulic transients in treatment systems.
3. Explain the reasons for system damage caused by transients.
4. Identify options for controlling or minimizing hydraulic transients.

Analyzing, designing, and troubleshooting steady-state hydraulic conditions in treatment systems are described in previous chapters. The possible error from inaccurate calculations or predictions may be a deviation of the actual flow rate from the design or expected flow rate. Although not being able to achieve the design flow rate at steady state in a system is not acceptable from an operational viewpoint, the presence of hydraulic transients in a system can result in catastrophic failure. Avoiding sudden failure of the system is an important topic for treatment system engineers.

Transients can arise from the following (Tullis 1989):

- events when valves are opened or closed,
- occurrences of pumps starting or stopping,
- cycling of pressure relief valves,
- operation of check valves, or
- other events that produce sudden changes in velocity in a portion of the system.

It is imperative that engineers have some insight into the transients that can occur in fluid systems and have an understanding of how transients may be controlled. Engineers need knowledge of the causes for hydraulic transients and of how to analyze for the presence and magnitude of certain transient events. The effect of transients on the system must also be understood, as the consequences of significant transients can be rupture and failure of system components. Viable control options may be needed.

Transients may be in the form of flow transients with relatively low-pressure "spikes," resulting from *unsteady flow*, or of potentially damaging *pressure waves*.

Unsteady Flow

Unsteady flow in a system may arise from starting pumps, opening valves, etc. Consider a system of two storage tanks with different water levels connected by a pipe, as shown in Fig. 11-1, where the flow is suddenly initiated. The flow rate in the pipe is a function of time (time after the flow is initiated) and will increase from zero to its maximum, steady-state value. First an equation describing the non-steady-state behavior of the water in the pipe will be developed (see also Wylie and Streeter 1982; Tullis 1989).

To resolve the behavior of the system, we must perform a momentum balance. The forces on the fluid in the pipe are depicted in Fig. 11-2.

The momentum balance derived in Chapter 4 was

$$\rho\left[\frac{\partial \vec{V}}{\partial t} + \left(V_x \frac{\partial(\vec{V})}{\partial x} + V_y \frac{\partial(\vec{V})}{\partial y} + V_z \frac{\partial(\vec{V})}{\partial z}\right)\right] = \sum \vec{f} \qquad (4\text{-}27)$$

where ρ is the fluid density, $\vec{V}$ is the fluid velocity, and $\vec{f}$ is the force per unit volume. Because $\partial(\vec{V})/\partial x = 0$, $\partial(\vec{V})/\partial y = 0$, and $\partial(\vec{V})/\partial z = 0$, the momentum balance equation becomes

$$\rho\frac{d\vec{V}}{dt} = \sum \vec{f} \qquad (11\text{-}1)$$

Since $\vec{f} = \vec{F}/\mathbf{v}$, we have

$$\rho\frac{d\vec{V}}{dt} = \sum \frac{\vec{F}}{\mathbf{v}} \qquad (11\text{-}2)$$

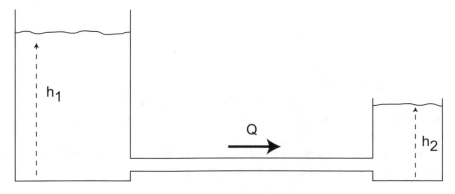

Figure 11-1. A system of two tanks connected by a pipe.

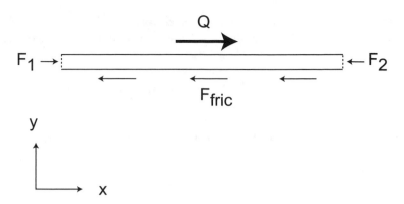

Figure 11-2. Forces on a fluid in a connecting pipe.

which may be simplified to

$$m\frac{d\vec{V}}{dt} = \sum \vec{F} \tag{11-3}$$

where m is mass.

We can resolve the forces in the x-direction. The force on left side of the fluid in the pipe from the static fluid head h_1 in the supply tank is

$$F_1 = P_1 \cdot A = \rho \cdot g \cdot h_1 \cdot A \tag{11-4}$$

where A is the cross-sectional area of the pipe. The force on right side of the fluid in the pipe from static fluid head h_2 in the discharge tank is

$$F_2 = P_2 \cdot A = \rho \cdot g \cdot h_2 \cdot A \tag{11-5}$$

The force on the fluid from friction, F_{fric}, is the pressure drop across the pipe multiplied by the pipe cross-sectional area:

$$F_{\text{fric}} = \Delta P \cdot A = 4f\rho\left(\frac{L}{D}\right)\left(\frac{V^2}{2}\right)A \tag{11-6}$$

where L and D are the length and diameter of the pipe respectively. Since $V = Q/A$, Eq. 11-6 can be modified to

$$F_{\text{fric}} = 4f\rho\left(\frac{L}{D}\right)\left(\frac{Q^2}{2 \cdot A}\right) \tag{11-7}$$

Substituting the equations that quantify each force on the fluid, Eqs. 11-4, 11-5, and 11-7, into Eq. 11-3 that was obtained from a momentum balance gives

$$(\rho \cdot L \cdot A)\frac{dV}{dt} = (\rho \cdot g \cdot h_1 A) - (\rho \cdot g \cdot h_2 A) - \left(4 \cdot f \cdot \rho \frac{L}{D} \frac{Q^2}{2 \cdot A} \right) \tag{11-8}$$

which, when simplified, yields a differential equation

$$\frac{dQ}{dt} = \frac{gA}{L}(h_1 - h_2) - \frac{4f}{D}\frac{Q^2}{2A} \tag{11-9}$$

Separating variables:

$$\frac{dQ}{\dfrac{gA}{L}(h_1 - h_2) - \dfrac{4f}{D}\dfrac{Q^2}{2A}} = dt \tag{11-10}$$

Integrating:

$$\int \frac{dQ}{2gDA^2(h_1 - h_2) - 4fLQ^2} = \int \frac{dt}{2ADL} \tag{11-11}$$

Yields (Griffin 2007):

$$\frac{\tanh^{-1}\left(\sqrt{\dfrac{2fL}{gDA^2(h_1 - h_2)}}\, Q \right)}{\sqrt{8fLgDA^2(h_1 - h_2)}} = \frac{t}{2ADL} + C \tag{11-12}$$

Rearranging:

$$\tanh^{-1}\left(\sqrt{\dfrac{2fL}{gDA^2(h_1 - h_2)}}\, Q \right) = \frac{\sqrt{2fg(h_1 - h_2)}}{DL}t + C \tag{11-13}$$

Solving for Q:

$$Q = \frac{\tanh\left(\sqrt{\dfrac{2fg(h_1 - h_2)}{DL}}\, t + C \right)}{\sqrt{\dfrac{2fL}{gDA^2(h_1 - h_2)}}} \tag{11-14}$$

At $t = 0$, $Q = 0$, so C must be equal to 0.

$$Q = \frac{\tanh\left(\sqrt{\dfrac{2fg(h_1 - h_2)}{DL}}\; t\right)}{\sqrt{\dfrac{2fL}{gDA^2(h_1 - h_2)}}} \tag{11-15}$$

This equation may be solved for specific conditions in the system (h_1, h_2, f, pipe diameter, and pipe length) to describe Q as a function of t. The friction factor is assumed to remain constant.

Example

How long does it take for the system depicted in Fig. 11-1 to reach 99% of steady-state flow rate, when $h_1 = 70$ m, $h_2 = 3$ m, $f = 0.0015$, the pipe diameter is 0.3 m, and the pipe length is 700 m?

Solution

Using Eq. 11-15, we find that the flow rate is constant after approximately 60 s. The steady state value is 0.685 m³/s (above 60 s).

By trial and error, evaluating Q as a function of time, it is found that it takes 27.2 s to reach 99% of the steady-state Q, which is 0.678 m³/s. Substituting values into Eq. 11-15:

$$Q = \frac{\tanh\left(\sqrt{\dfrac{2 \cdot 0.0015 \cdot 9.81\dfrac{m}{s^2} \cdot (70\ m - 3\ m)}{0.3\ m \cdot 700\ m}} \cdot 27.2s\right)}{\sqrt{\dfrac{2 \cdot 0.0015 \cdot 700\ m}{9.81\dfrac{m}{s^2} \cdot 0.3\ m \cdot \left(\dfrac{\pi}{4}(0.3\ m)^2\right)^2 (70\ m - 3\ m)}}} = 0.678\ \frac{m^3}{s}$$

So it takes about 27 s to reach steady state in this system. By the same analysis, a 2,000-m-long pipe with an order of magnitude lower friction factor takes approximately 133 s to reach steady state—a significant increase in time. Any change in the system that will upset steady state, such as changing control valve settings or turning a pump on, should be allowed this amount of time as a rough approximation for the flow to reach steady state again.

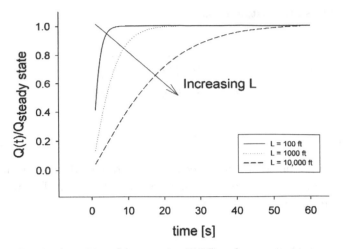

Figure 11-3. Transient response with increasing pipe length. The parameters used are $f = 0.0015$, $\Delta h = 200$ ft (61 m), and pipe diameter = 12 in. (~30 cm).

Trends with increasing pipe length, head difference, pipe diameter, and friction factor are shown in Figs. 11-3 through 11-6.

Pressure Waves

Consider a horizontal pipe with frictionless flow from a tank to a valve as shown in Fig. 11-7. If the valve is instantaneously adjusted to the partially closed position,

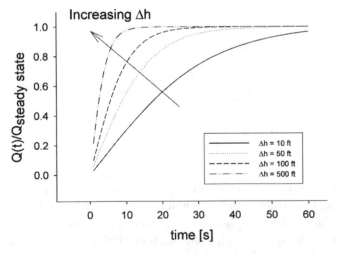

Figure 11-4. Transient response with increasing static height (head) difference. The parameters used are $f = 0.0015$, $L = 1,000$ ft (~300 m), and pipe diameter = 12 in. (~30 cm).

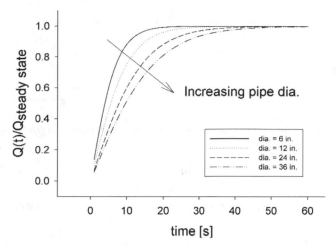

Figure 11-5. Transient response with increasing pipe diameter. The parameters used are $f = 0.0015$, $L = 1,000$ ft ($\sim$300 m), and $\Delta h = 100$ ft ($\sim$30 m).

the velocity in the system will be decreased by ΔV. This decrease in velocity by ΔV is manifested as an increase in pressure by ΔP. This increase in pressure resulting from the velocity decrease should be obvious from Bernoulli's equation. As the valve was instantaneously adjusted, the pressure increase happens instantaneously as well, and a pressure wave forms in the pipe. This pressure wave, and its consequences, is known as *water hammer*. The pressure wave compresses the liquid in the system, expanding the conduit that the liquid is in. The pressure wave travels back up the pipe opposite to the direction of flow with a wave velocity in the liquid of a.

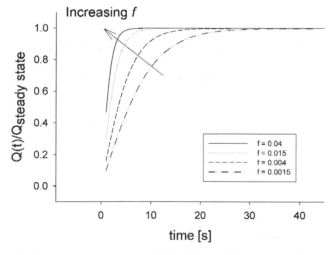

Figure 11-6. Transient response with increasing friction factor. The parameters used are pipe diameter = 12 in. ($\sim$30 cm), $L = 1,000$ ft ($\sim$300 m), and $\Delta h = 100$ ft ($\sim$30 m).

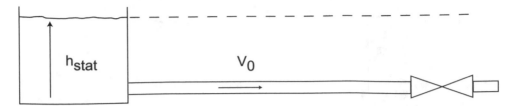

Figure 11-7. System composed of a horizontal tank, pipe, and control valve.

So, on instantaneous valve closure, the pressure wave is as shown in Fig. 11-8. A control volume is drawn around the pressure wave in the pipe. See Fig. 11-9.

The word equation for momentum balance (derived in Chapter 4) is

$$\begin{pmatrix} \text{accumulation} \\ \text{(rate of change of} \\ \text{momentum in system)} \end{pmatrix} = \begin{pmatrix} \text{net rate of flow of} \\ \text{momentum through} \\ \text{system boundaries} \end{pmatrix} + \begin{pmatrix} \text{sum of forces on} \\ \text{fluid boundaries} \end{pmatrix} \qquad (4\text{-}29)$$

The first term for the rate of change of momentum in the control volume is

$$[\rho A(a - V_0)\Delta t] \cdot \left[\frac{\Delta V}{\Delta t} \right] \qquad (11\text{-}16)$$

where V_0 is the initial velocity and ΔV is the change in velocity in the control volume. The term in the first set of brackets is the mass of fluid affected, and the term in the second set of brackets is the velocity change over the time interval, Δt. Equation 11-16 can be simplified to

$$\rho A(a - V_0)\Delta V \qquad (11\text{-}17)$$

The net momentum flux (into and out of the control volume) is

$$\rho \cdot A(V_0 + \Delta V)^2 - \rho \cdot A \cdot V_0^2 \qquad (11\text{-}18)$$

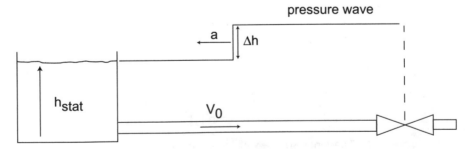

Figure 11-8. System composed of a horizontal tank, pipe, and control valve when the valve is closed. *Source:* Adapted from Tullis (1989).

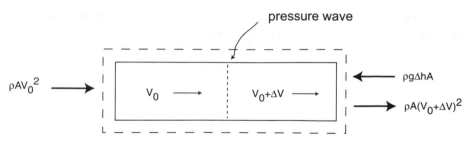

Figure 11-9. Control volume around a pressure wave showing the forces on the control volume. *Source:* Adapted from Wylie and Streeter (1982) and Tullis (1989).

Expanding the first term gives

$$\rho \cdot A[V_0^2 + 2V_0\Delta V + \Delta V^2] - \rho \cdot A \cdot V_0^2 \qquad (11\text{-}19)$$

which simplifies to

$$\rho \cdot A[2V_0\Delta V + \Delta V^2] \qquad (11\text{-}20)$$

Since $V_0\Delta V \gg \Delta V^2$ we can drop the second term to get

$$2\rho \cdot A \cdot V_0 \cdot \Delta V \qquad (11\text{-}21)$$

The net force on the fluid in the control volume is

$$-\rho{\cdot}g \cdot \Delta h \cdot A \qquad (11\text{-}22)$$

Substituting Eq. 11-17 for the rate of change of momentum, Eq. 11-21 for the net momentum flux, and Eq. 11-22 for the net force into the momentum balance gives

$$-\rho g \Delta h A = 2\rho A V_0 \Delta V + \rho A(a - V_0)\Delta V \qquad (11\text{-}23)$$

Solving for Δh yields

$$\Delta h = \frac{a\Delta V}{g}\left(1 + \frac{V_0}{a}\right) \qquad (11\text{-}24)$$

Because a is usually much greater than V_0, $V_0/a \approx 0$, and so

$$\Delta h = -\frac{a\Delta V}{g} \qquad (11\text{-}25)$$

Because the pressure waves can be superimposed, the increased head (pressure) from multiple pressure waves can be quantified with

$$\sum \Delta h = -\frac{\sum a \Delta V}{g} \qquad (11\text{-}26)$$

Note that this derivation follows that by Tullis (1989), where more details may be found.

Example

A pipe is carrying 10 °C water at 4.5 m/s. The flow is immediately throttled to 2 m/s. If the pressure wave speed is 1,000 m/s, what is the pressure increase in pascals?

Solution

The change in pressure head is

$$\Delta h = -\frac{a \Delta V}{g} = -\frac{1{,}000\,\dfrac{\text{m}}{\text{s}} \cdot \left(2 - 4.5\,\dfrac{\text{m}}{\text{s}}\right)}{9.8\,\dfrac{\text{m}}{\text{s}^2}} = 255 \text{ m}$$

which gives an increase in pressure of

$$\Delta P = \rho \cdot g \cdot \Delta h = 999.7\,\frac{\text{kg}}{\text{m}^3} \cdot 9.8\,\frac{\text{m}}{\text{s}^2} \cdot 255 \text{ m}$$

$$= 2.5 \times 10^6\,\frac{\text{kg}}{\text{m} \cdot \text{s}^2} = 2.5 \times 10^6 \text{ Pa}$$

Velocity of the Pressure Wave

The head increase or pressure increase in a pipe resulting from pressure wave propagation depends on the velocity of the pressure wave. In the previous example, the pressure wave speed was assumed, but an expression for the pressure wave speed can be derived. When an event creates a pressure wave in a pipe (such as a valve slamming closed), the pressure wave travels down the pipe. As the wave travels along, the pipe expands and the fluid compresses slightly. A mass balance can be performed to arrive at the wave speed. The derivation is adapted from Tullis (1989).

The total mass entering the pipe between $t = 0$ and $t = L/a$ (the time for the pressure wave to travel the length L of the pipe) is

$$\rho A V_0 t \tag{11-27}$$

or, since $t = L/a$,

$$\rho A V_0 \frac{L}{a} \tag{11-28}$$

Because the pipe expands, there is an increase in the mass of liquid stored in the pipe. The increase is quantified with

$$\rho L \Delta A \tag{11-29}$$

There is also an increase in the mass of liquid stored in the pipe owing to fluid compression:

$$L A \Delta \rho \tag{11-30}$$

Combining into a mass balance we have

$$\rho A V_0 \frac{L}{a} = \rho L \Delta A + L A \Delta \rho \tag{11-31}$$

Rearranging and substituting in $g \Delta h / a$ for velocity gives

$$a^2 = \frac{g \Delta h}{\dfrac{\Delta A}{A} + \dfrac{\Delta \rho}{\rho}} \tag{11-32}$$

The liquid bulk modulus is defined as $K = \Delta P / (\Delta \rho / \rho)$ and $\Delta P = \rho g \Delta h$, so Eq. 11-32 becomes

$$a^2 = \frac{\dfrac{K}{\rho}}{1 + \dfrac{K \Delta A}{A \Delta P}} \tag{11-33}$$

When a pipe expands, the change in area can be defined as

$$\Delta A = \text{circumference} \cdot \text{change in radius} \tag{11-34}$$

From solid mechanics, the change in pipe radius with pressure change is

$$\frac{\text{change in}}{\text{radius}} = \frac{r^2 \cdot \Delta P}{x_{\text{wall}} \cdot E} \tag{11-35}$$

where r is the pipe radius, x_{wall} is the thickness of the pipe wall, and E is Young's modulus of elasticity. So Eq. 11-34 becomes

$$\Delta A = (\pi \cdot d) \cdot \left(\frac{d^2 \cdot \Delta P}{4 \cdot x_{wall} \cdot E} \right) \qquad (11\text{-}36)$$

where d is the pipe diameter. Rearranging gives

$$\Delta A = \left(\frac{\pi d^2}{4} \right) \cdot \left(\frac{\Delta P \cdot d}{x_{wall} \cdot E} \right) = \frac{A \cdot \Delta P \cdot d}{x_{wall} \cdot E} \qquad (11\text{-}37)$$

Substituting into Eq. 11-33 then yields

$$a = \frac{\sqrt{\dfrac{K}{\rho}}}{\sqrt{1 + \dfrac{K \cdot d}{x_{wall} \cdot E}}} \qquad (11\text{-}38)$$

It has been found that the pipe anchors affect how, and to what extent, the pipe responds to pressure waves. So the pressure wave speed is affected as well. This is accounted for by adding an anchor coefficient C in Eq. 11-38 (Tullis 1989) as follows:

$$a = \frac{\sqrt{\dfrac{K}{\rho}}}{\sqrt{1 + C \dfrac{K \cdot d}{x_{wall} \cdot E}}} \qquad (11\text{-}39)$$

Anchor coefficients are listed in Table 11-1 and piping material properties are listed in Table 11-2.

Therefore pressure wave speed, as well as the ultimate maximum head and pressure to which piping components are subjected, is a function of the bulk liquid

Table 11-1. Anchor coefficients for calculating the pressure wave speed in a pipe.
Source: Tullis (1989).

Anchor configuration	Anchor coefficient, C
Anchors with expansion joints	1
Anchor on upstream end of pipe	$1 - 0.5\,\mu$[a]
Anchors preventing any movement along the axis of the pipe	$1 - \mu^2$

[a]μ is Poisson's ratio for the pipe material.

Table 11-2. Piping material properties.

Material	E [GPa] (psi)	Poisson's ratio, μ
Cast steel	197 (28.5×10^6)	0.265
Cold-rolled steel	203 (29.5×10^6)	0.287
Stainless steel	190 (27.6×10^6)	0.305
Cast iron	93.0–145 $[(13.5–21.0) \times 10^6]$	0.211–0.299
Copper	108 (15.6×10^6)	0.355
Monel	172 (25.0×10^6)	0.315
Inconel	214 (31.0×10^6)	0.270–0.380
Titanium	103–110 $[(15–16) \times 10^6]$	0.34
ABS	0.90–2.9 $[(1.3–4.2) \times 10^5]$	
PVC	2.4–4.1 $[(3.5–6.0) \times 10^5]$	0.45
CPVC	2.3-3.2 $[(3.4–4.7) \times 10^5]$	
Concrete	21–34 $[(3–5) \times 10^6]$	0.30

Source: Data are from Avallone and Baumeister (1996), except Poisson ratios for PVC and concrete, which are from Tullis (1989).

modulus, the liquid density, Young's modulus of elasticity of the pipe material, the diameter and wall thickness of the pipe, and the anchor coefficients.

Example

Calculate the wave speed of 45 °F water flowing at 10 ft/s in 6-in.-diameter cast iron pipe that is anchored along the axis of the pipe. The wall thickness of the pipe is 0.280 in. Assume the bulk modulus of water is 3.2×10^5 psi.

Solution

The water density at 45 °F is 62.421 lbm/ft^3. The average modulus of elasticity of cast iron pipe in the range listed in Table 11-2 is 17.2×10^6 psi, which is as good an estimate as available. Because the pipe is anchored along the axis, the anchor coefficient is calculated with

$$C = 1 - \mu^2$$

Again, taking Poisson's ratio to be the average of the range of values listed in Table 11-2, or 0.255, we have

$$C = 1 - 0.255^2 = 0.935$$

Calculating the pressure wave speed with Eq. 11-39 gives

$$a = \frac{\sqrt{\dfrac{K}{\rho}}}{\sqrt{1 + C\dfrac{K \cdot d}{x_{\text{wall}} \cdot E}}} = \frac{\sqrt{\dfrac{3.2 \times 10^5 \dfrac{\text{lbf}}{\text{in.}^2}}{62.421 \dfrac{\text{lbm}}{\text{ft}^3}} \cdot \left(\dfrac{144 \text{ in.}^2}{1 \text{ ft}^2} \right) \cdot \left(32.2 \dfrac{\text{ft} \cdot \text{lbm}}{\text{lbf} \cdot \text{s}^2} \right)}}{\sqrt{1 + \left[0.935 \left(\dfrac{3.2 \times 10^5 \text{ psi} \cdot (6 \text{ in.})}{(0.280 \text{ in.}) \cdot 17.2 \times 10^6 \text{ psi}} \right) \right]}}$$

$$= 4{,}160 \text{ ft/s (or } \sim 1{,}270 \text{ m/s)}$$

Note the inclusion of g_c!

The head increase in the pipe owing to the pressure wave can be calculated with Eq. 11-25:

$$\Delta h = -\frac{a \Delta V}{g} = -\frac{4{,}160 \text{ ft/s} \cdot (0 - 10 \text{ ft/s})}{32.17 \text{ ft/s}^2} = -\frac{4{,}160 \text{ ft/s} \cdot (0 - 10 \text{ ft/s})}{32.17 \text{ ft/s}^2}$$

$$= 1{,}293 \text{ ft (or } 394 \text{ m)}$$

The pressure increase or "spike" is therefore

$$\Delta P = \rho \cdot g \cdot \Delta h = 62.42 \frac{\text{lbm}}{\text{ft}^3} \cdot 32.17 \text{ ft/s}^2 \cdot 1{,}293 \text{ ft} \cdot \frac{1 \text{ lbf} \cdot \text{s}^2}{32.2 \text{ ft} \cdot \text{lbm}} \cdot \frac{1 \text{ ft}^2}{144 \text{ in.}^2}$$

$$= 561 \text{ psid (or } 3.87 \text{ kPa)}$$

So slamming a valve shut with a fluid velocity of this magnitude (10 ft/s) produces a 561 psi pressure spike. If the system components cannot take this pressure, catastrophic rupture may take place.

Now consider this scenario taking place in a concrete pipe. The pressure wave speed would be

$$a = \frac{\sqrt{\dfrac{3.2\times10^5 \text{ lbf}/\text{in}^2}{62.421 \text{ lbm}/\text{ft}^3}\cdot\left(\dfrac{144 \text{ in.}^2}{1 \text{ ft}^2}\right)\cdot\left(32.2\dfrac{\text{ft}\cdot\text{lbm}}{\text{lbf}\cdot\text{s}^2}\right)}}{\sqrt{1+\left[0.935\left(\dfrac{3.2\times10^5 \text{ psi}\cdot(6 \text{ in.})}{(0.280 \text{ in.})\cdot 4\times10^6 \text{ psi}}\right)\right]}} = 3,050 \text{ ft/s (or } \sim 930 \text{ m/s)}$$

resulting in a pressure head spike of

$$\Delta h = -\frac{3,050 \text{ ft/s}\cdot(0-10 \text{ ft/s})}{32.17 \text{ ft/s}^2} = 947 \text{ ft (or 289 m)}$$

which is equivalent to a pressure spike of 410 psid (or ~2.8 kPa). If the identical conditions (valve slamming shut with water flowing at 10 ft/s) occur in a rubber pipe, assuming $E = 0.01 \times 10^6$ psi (Wylie and Streeter 1982) and $\mu = 0.5$, we would get

$$a = \frac{\sqrt{\dfrac{3.2\times10^5 \text{ lbf}/\text{in.}^2}{62.421 \text{ lbm}/\text{ft}^3}\cdot\left(\dfrac{144 \text{ in.}^2}{1 \text{ ft}^2}\right)\cdot\left(32.2\dfrac{\text{ft}\cdot\text{lbm}}{\text{lbf}\cdot\text{s}^2}\right)}}{\sqrt{1+\left[0.750\left(\dfrac{3.2\times10^5 \text{ psi}\cdot(6 \text{ in.})}{(0.280 \text{ in.})\cdot 0.01\times10^6 \text{ psi}}\right)\right]}} = 215 \text{ ft/s (or 65.5 m/s}$$

$$\Delta h = \frac{215 \text{ ft/s}\cdot(0-10 \text{ ft/s})}{32.17 \text{ ft/s}^2} = 374 \text{ ft (or 114 m)}$$

This would give a pressure spike of 29 psid (or 200 kPa), which is significantly lower than the pressure spikes for more rigid materials.

Minimizing Occurrence or Damage from Transients

The pressure spikes in treatment systems from these pressure waves may be prevented by eliminating the cause, the velocity change that precedes the pressure wave. By reducing the fluid velocity in the system, the difference in velocity will be smaller when a valve instantly closes than for a system with high velocity. By minimizing velocity change, the pressure wave has a lower magnitude, as predicted by Eq. 11-25. The time for a given velocity change in a system can be lengthened, thereby "softening" the pressure wave. This can be done by providing for longer valve closure times so that an instantaneous change in velocity does not occur. Pumps can also be controlled so that the difference in system velocity arising from a pump starting up or stopping happens over a period of time, not all at once.

The pressure arising from pressure wave transients can be controlled through the use of pressure relief valves; these will prevent rupture of the components in the system (if the relief valve is properly applied). The presence of air in the system, which is much more compressible than liquids, will help mitigate the effects of a pressure wave. Either air can be injected into the liquid or an air chamber with or without a bladder separating the liquid from the air may be employed. The pressure in the system is reduced by the air, as energy can be dissipated in compression of the air, reducing the pressure wave speed. Also, as illustrated in the example with the rubber pipe, a more flexible component in the system can reduce the pressure wave velocity, and thereby the pressure spike for a given instantaneous velocity change. So either flexible rubber hoses can be installed in the system or piping material with lower Young's modulus of elasticity can be used.

Non-steady-state surges can be mitigated with surge tanks or stand pipes. A large stand pipe to control surge is shown in Fig. 11-10.

Figure 11-10. A stand pipe used to control surge in a piping system from a reservoir.

Symbol List

a	pressure wave velocity in the liquid
A	cross-sectional area
C	anchor coefficient
D	diameter
E	Young's modulus of elasticity
f	force per unit volume
F	force
g	acceleration due to gravity
h	static head
K	liquid bulk modulus
L	length
m	mass
P	pressure
Q	volumetric flow rate
t	time
V	velocity
v	volume
ρ	fluid density

Problems

1. A valve in a treatment system is opened from partially throttled to full, increasing the water velocity from 1.5 to 5 m/s. The water is at 15 °C. What is the magnitude of the pressure spike in pascals if $a = 1,200$ m/s?

2. Water is standing still in an 8-in.-diameter 400-ft-long pipe until a valve is opened. The flow rate in the pipe starts from zero and approaches steady state with a constant $\Delta h = 50$ ft across the pipe. If there is no friction, what is the flow rate in gallons per minute in the pipe after 20 s?

3. What is the pressure spike in pascals resulting from an instantaneous valve closing with a water flow of 3 m/s in a 15-cm-diameter pipe? The water is at 8 °C. The pipe is CPVC and is anchored at the upstream end. The wall thickness of the pipe is 0.711 cm. Assume that $\mu = 0.45$ and that the bulk modulus of water is 2.2×10^9 Pa. How does the pressure spike compare to those in the examples in this chapter.

4. A 4-in.-diameter pipe is to be used to transfer 60 °F water at 12 ft/s. It is anticipated that the flow will be intermittent with a control valve providing instant closure at times. The pipe wall will be 0.237 in. thick, but the material must be chosen to keep the pressure spike below 250 psi to avoid rupture of components in the system. The pipe will be anchored along its axis. Will the use of steel pipe keep the pressure spike below the allowable pressure? What pipe materials are acceptable?

5. You are the engineer at a water treatment plant who is troubleshooting a "water hammer" pressure wave problem that occurs in a pipe when a valve is closed. Specifically, what are your options to eliminate this pressure wave trouble or mitigate the damage from it?

References

Avallone, E. A., and Baumeister, T., III, eds. (1996). *Marks' Standard Handbook for Mechanical Engineers*, McGraw-Hill, New York.

Griffin, Robert, personal communication, December 9, 2007.

Tullis, J. P. (1989). *Hydraulics of Pipelines*, Wiley, New York.

Wylie, E. B., and Streeter, V. L. (1982). *Fluid Transients*, FEB Press, Ann Arbor, MI.

CHAPTER 12

Open Channel Flow

Chapter Objectives

1. Summarize the theoretical basis for Manning's equation, and use it to quantify flow rate in open channel flow.
2. Examine depth–energy curves in open channel flow.
3. Classify supercritical and subcritical flow in open channels.

Open channel flow is gravity-induced liquid flow in which there is a free surface. Typically, the top surface is open to the atmosphere either in a partially full pipe or in a channel. Open channels are used in many treatment systems to convey water or wastewater; therefore, it is important for engineers to understand how flow in open channels behaves. An open channel of flowing water is shown in Fig. 12-1.

The Manning Equation

One of the most useful equations for determining flow in open channels is the Manning equation. Flow in open channels may be *uniform* as depicted in Fig. 12-2 or *nonuniform*. The velocity and depth are uniform along the length of the control volume for *uniform flow* (i.e., $V_1 = V_2$) at liquid depth y. Nonuniform flow is characterized by varying or nonuniform velocities and depths along the channel length.

The forces acting on a control volume of fluid in Fig. 12-2 are as follows:

- The frictional force is

$$F_f = -\tau_0 \cdot P \cdot L \qquad (12\text{-}1)$$

where τ_0 is the shear stress at the wall, P is the wetted perimeter, and L is the length of the control volume.

Figure 12-1. Open channel flow at a wastewater treatment plant.

- The force of gravity in the direction of flow is

$$F_g = \gamma \cdot A \cdot L \sin \theta \qquad (12\text{-}2)$$

where γ is the specific weight of the fluid, A is the cross-sectional area, L is the length of the control volume, and θ is the angle of inclination of the channel bottom.

- Hydrostatic forces are the forces attributed to the height of the liquid: F_1 at the left side of the fluid element and F_2 at the right side. For uniform flow the hydrostatic forces, $F_1 = F_2$, cancel out.

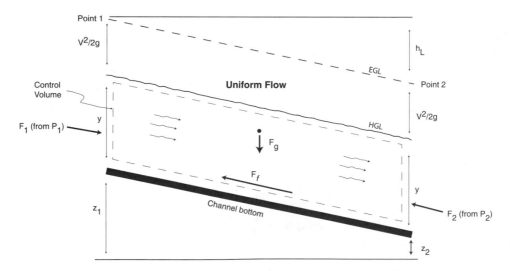

Figure 12-2. Uniform open channel flow. *Source:* Adapted from Mays (2000).

Summing forces (in the direction of flow) gives

$$F_f + F_g + (F_1 - F_2) = 0 \tag{12-3}$$

For uniform flow ($F_1 = F_2$), substituting F_f and F_g from Eqs. 12-1 and 12-2 gives

$$(-\tau_0 \cdot P \cdot L) + (\gamma \cdot A \cdot L \cdot \sin \theta) = 0 \tag{12-4}$$

For small angles of inclinations the slope of the channel bottom is $S_0 \approx \sin \theta$. Rearranging gives

$$\tau_0 = \frac{\gamma \cdot A \cdot L \cdot S_0}{P \cdot L} \tag{12-5}$$

and substituting the hydraulic radius R for A/P yields

$$\tau_0 = \gamma \cdot R \cdot S_0 \tag{12-6}$$

For turbulent flow, as for closed-conduit flow, we can define the shear stress as a function of the fluid density ρ, fluid velocity V, and a coefficient C_f characterizing the resistance to flow:

$$\tau_0 = C_f \cdot \rho \cdot \left(\frac{V^2}{2} \right) \tag{12-7}$$

Combining the Eqs. 12-6 and 12-7 results in

$$\gamma \cdot R \cdot S_0 = C_f \cdot \rho \cdot \left(\frac{V^2}{2} \right) \tag{12-8}$$

Rearranging to solve for velocity gives

$$V = \sqrt{\frac{2g}{C_f}} \cdot \sqrt{R \cdot S_0} \tag{12-9}$$

Redefining the constant $\sqrt{2g/C_f}$ as simply C produces

$$V = C \cdot \sqrt{RS_0} \tag{12-10}$$

which is called the *Chezy* equation, where C is the Chezy coefficient.

In the late 1800s, Robert Manning used empirical data to develop an expression to calculate C from roughness coefficients (n) and the hydraulic radius of the channel (R). The expression is empirical and is as follows:

$$C = \frac{1}{n} R^{1/6} \tag{12-11}$$

Substituting Manning's expression for C into the Chezy equation yields the *Manning equation*, which describes the velocity V (in meters per second) in a channel as a function of a roughness coefficient n, the hydraulic radius R (in meters), and the slope of the channel bottom, S (in meters per meter):

$$V = \frac{1}{n} \cdot R^{2/3} \cdot \sqrt{S_0} \qquad \text{(SI units)} \qquad (12\text{-}12)$$

Common ranges of values to use for Manning's roughness coefficient n are listed in Table 12-1.

Manning's equation can also be written in SI units in terms of flow rate Q (since $Q = AV$):

$$Q = \frac{1}{n} \cdot A \cdot R^{2/3} \cdot \sqrt{S_0} \qquad \text{(SI units)} \qquad (12\text{-}13)$$

where Q is in cubic meters per second and A is in square meters. In U.S. Customary units,

$$V = \frac{1.49}{n} \cdot R^{2/3} \cdot \sqrt{S_0} \qquad \text{(U.S. Customary units)} \qquad (12\text{-}14)$$

$$Q = \frac{1.49}{n} \cdot A \cdot R^{2/3} \cdot \sqrt{S_0} \qquad \text{(U.S. Customary units)} \qquad (12\text{-}15)$$

where V is in feet per second, R is in feet, Q is in cubic feet per second, and A is in square feet. Manning's equation was derived under the assumption of turbulent

Table 12-1. Values for Manning's roughness coefficient.

	Manning's roughness coefficient, n		
Material	*Minimum*	*Average*	*Maximum*
Smooth brass	0.009	0.010	0.013
Welded steel	0.010	0.012	0.014
Riveted, spiral steel	0.013	0.016	0.017
Coated cast iron	0.010	0.013	0.014
Uncoated cast iron	0.011	0.014	0.016
Black wrought iron	0.012	0.014	0.015
Galvanized wrought iron	0.013	0.016	0.017
Glass	0.009	0.010	0.013
Concrete, finished	0.011	0.012	0.014
Concrete, unfinished smooth wood form	0.012	0.014	0.016
Clay, vitrified sewer	0.011	0.014	0.017
Sanitary sewers coated with biological slime	0.012	0.013	0.016

Source: Data from Chow 1959.

Rectangle

Triangle

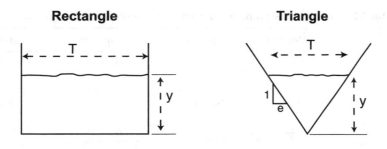

Trapezoid

Circle

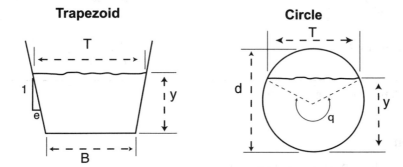

Figure 12-3. Common cross sections for open channels. *Source:* Adapted from Chow 1959.

flow (see the previous expression for τ_0). The conditions for turbulent flow in open channels occur when (Mays 2000)

$$n^6\sqrt{RS_f} \geq 1.1 \times 10^{-13} \qquad \text{(SI units)} \qquad (12\text{-}16)$$

$$n^6\sqrt{RS_f} \geq 1.9 \times 10^{-13} \qquad \text{(U.S. Customary units)} \qquad (12\text{-}17)$$

where S_f is the friction head loss, h_L, per unit channel length (see Fig. 12-1). Note that $S_f = S_0$ for uniform flow.

Common cross sections that are used for open channel conduits are shown in Fig. 12-3. The geometrical properties of these cross sections are listed in Table 12-2.

Example

What is the flow rate through a concrete rectangular channel 2.4 m wide, with a bottom slope of 0.0005 m/m and a flow depth of 0.8 m? The concrete is unfinished, as left by the forms.

Table 12-2. Geometrical properties of common open channel cross sections.

	Rectangle	Triangle	Trapezoid	Circular
Area	$y \cdot T$	$y^2 \cdot e$	$y(B + ey)$	$\frac{1}{8}(\theta - \sin \theta)d^2$
Wetted perimeter	$T + 2y$	$2y\sqrt{1 + e^2}$	$B + 2y\sqrt{1 + e^2}$	$\frac{1}{2}\theta d$
Hydraulic radius, R	$\dfrac{T \cdot y}{T + 2y}$	$\dfrac{ey}{2\sqrt{1 + e^2}}$	$\dfrac{(B + ey)y}{B + 2y\sqrt{1 + e^2}}$	$\frac{1}{4}\left(1 - \dfrac{\sin \theta}{\theta}\right)d$
Top width, T	T	$2ey$	$B + 2ey$	$d \cdot \sin\left(\dfrac{\theta}{2}\right)$
Hydraulic depth, D	y	$\frac{1}{2}y$	$\dfrac{(B + ey)y}{B + 2ey}$	$\left(\dfrac{\theta - \sin \theta}{\sin \frac{1}{2}\theta}\right)\dfrac{d}{8}$

Source: Adapted from Chow (1959).

Solution

The hydraulic radius is calculated with

$$R = \frac{T \cdot y}{T + 2y} = \frac{2.4 \text{ m} \cdot 0.8 \text{ m}}{2.4 \text{ m} + (2 \cdot 0.8 \text{ m})} = 0.48 \text{ m}$$

The cross-sectional area is

$$A = T \cdot y = 2.4 \text{ m} \cdot 0.8 \text{ m} = 1.92 \text{ m}^2$$

For unfinished concrete, $n = 0.014$ from Table 12-1 (where an average value is assumed). Using Manning's equation to calculate the flow rate, we have

$$Q = \frac{1}{n} \cdot A \cdot R^{2/3} \cdot \sqrt{S_0} = \frac{1}{0.014} \cdot 1.92 \text{ m}^2 \cdot (0.48 \text{ m})^{2/3} \cdot \sqrt{0.0005 \text{ m/m}}$$

$$= 1.88 \text{ m}^3/\text{s (or} \sim 66 \text{ ft}^3/\text{s)}$$

Specific Energy of Open Channel Flow

In an earlier chapter, an energy balance (Bernoulli's equation) was derived:

$$\Delta\left(\frac{P}{\rho g} + z + \frac{V^2}{2g}\right) = \frac{-dW}{gdm} - \frac{\Im}{g} \tag{12-18}$$

For flow in an open channel without work done by or to the fluid (i.e., there is no pump), and in the absence of friction for the moment, the energy at any point in the flow is

$$E = \frac{P}{\rho g} + z + \frac{V^2}{2g} \qquad (12\text{-}19)$$

where z is the height of the channel bottom (see Fig. 12-2), and E is the total energy at that point in the fluid, which is the total head. The pressure P is a function of the depth y of the fluid at that point:

$$P = \rho g y \qquad (12\text{-}20)$$

Substituting Eq. 12-20 for P gives

$$E = \frac{\rho g y}{\rho g} + z + \frac{V^2}{2g} \qquad (12\text{-}21)$$

Canceling ρ and g from the second term in the equation gives the energy equation

$$E = y + z + \frac{V^2}{2g} \qquad (12\text{-}22)$$

If we set the bottom of the channel as the datum, $z = 0$, then

$$E = y + \frac{V^2}{2g} \qquad (12\text{-}23)$$

Equation 12-23 defines the *specific energy*, or the total head of the liquid above the bottom of the channel.

Rewriting Eq. 12-23 in terms of flow rate (using $Q = VA$) gives

$$E = y + \frac{Q^2}{2gA^2} \qquad (12\text{-}24)$$

For a liquid flowing at a given flow rate Q, through a channel of a given cross-sectional (wetted) area A, the liquid must have a certain flow depth y. The specific energy can be calculated with this equation and plotted for increasing y. See Fig. 12-4.

From Fig. 12-4, it can be seen that there is a minimum in the energy curve at $dE/dy = 0$. This minimum occurs at *critical flow*. The liquid depth associated with critical flow is called the *critical depth*. If the flow in the channel has a flow depth greater than the critical depth, the flow has a greater energy than at critical flow, and the energy predominantly results from the liquid depth (head, y). A flow depth

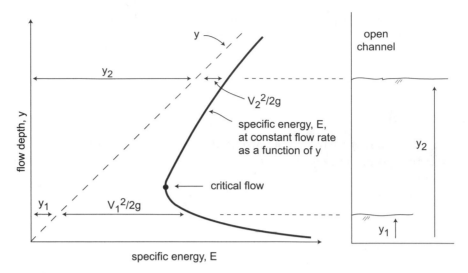

Figure 12-4. Graph of specific energy as a function of flow depth in an open channel.

less than the critical depth again has a greater energy than at critical flow, but the energy is predominantly in the form of kinetic energy owing to the increased velocity.

At the point of critical flow, the slope (dE/dy) is zero:

$$\frac{dE}{dy} = 0 \tag{12-25}$$

Taking the derivative of E in Eq. 12-24 gives

$$\frac{dE}{dy} = 1 - \frac{Q^2}{gA^3}\frac{dA}{dy} = 0 \tag{12-26}$$

The width of the channel at the top surface is defined as T. The value for T is a function of the cross-sectional area and depth: $T = dA/dy$. Substituting this definition for T into Eq. 12-26 gives

$$1 - \frac{TQ^2}{gA^3} = 0 \tag{12-27}$$

Rearranging then yields

$$\frac{TQ^2}{gA^3} = 1 \tag{12-28}$$

Since this derivation is for conditions at critical flow, we can designate T_c, A_c, Q_c, and V_c at critical flow and so

$$\frac{T_c Q_c^2}{g A_c^3} = 1 \tag{12-29}$$

Since $Q_c = V_c/A_c$ we have

$$\frac{T_c \left(\dfrac{V_c}{A_c}\right)^2}{g A_c^3} = 1 \tag{12-30}$$

Rearranging gives

$$\frac{V_c^2}{g \left(\dfrac{A_c}{T_c}\right)} = 1 \tag{12-31}$$

Defining the hydraulic depth at critical flow, D_c, as A_c/T_c we then arrive at

$$\frac{V_c^2}{g D_c} = 1 \tag{12-32}$$

Taking the square root of both sides yields

$$\frac{V_c}{\sqrt{g D_c}} = 1 \tag{12-33}$$

The first term is a form of the *Froude number*

$$\text{Fr} = \frac{V}{\sqrt{g D}} \tag{12-34}$$

which is equal to 1 at critical flow. The *Froude number* is a dimensionless parameter that is the ratio of inertial force to gravitational force (Avallone and Baumeister 1996). From a dynamic similarity approach, the inertia force of a fluid element divided by the gravitational force is

$$\text{Fr} \propto \frac{\text{inertial force}}{\text{gravitational force}} = \frac{\text{mass} \cdot \text{acceleration}}{\text{mass} \cdot \text{gravitational acc.}} = \frac{(m)\left(\dfrac{L}{T^2}\right)}{(m)g} \tag{12-35}$$

Cancelling m from the numerator and denominator, and multiplying by L/L:

$$\text{Fr} \propto \frac{\left(\dfrac{L}{T^2}\right)}{g}\frac{L}{L} = \frac{\left(\dfrac{L^2}{T^2}\right)}{Lg} = \frac{V^2}{Lg} \tag{12-36}$$

Taking the square root:

$$\text{Fr} \propto \frac{V}{\sqrt{Lg}} \tag{12-37}$$

Therefore the Froude number is a function of a characteristic length L, a characteristic velocity V, and g, the acceleration due to gravity. For the open channel Froude number, the characteristic velocity is defined as the *flow velocity* in the channel, and the characteristic depth is taken to be the *hydraulic depth* of the flow in the channel as called out in Eq. 12-34.

We can observe from Fig. 12-4 that when the hydraulic depth is greater than the critical flow depth, the velocity is lower than that at critical depth. This is termed *subcritical flow*. If $\text{Fr} = 1$ at critical flow, Fr must be less than 1 for subcritical flow (as V is smaller and D is greater). For supercritical flow, the converse is true: Fr must be greater than 1. See Table 12.3 and Fig. 12-5.

When flow through an open channel is near the critical flow condition, small changes in specific energy can cause large changes in flow depth, as shown in Fig. 12-6. This is easily explained by the specific energy curve in Fig. 12-5. The slope near the critical flow state is very high (dy/dE). So a small change in energy (ΔE) produces a large change in flow depth (Δy) near the critical state because of this high slope. For hydraulic jumps, as the flow goes from supercritical, through critical flow, and into subcritical flow, energy is dissipated by turbulence, and an increase in flow depth is observed.

It is common for sewers that transport raw sewage to be designed for open channel flow. For circular cross sections, an actual flow depth greater than about 90% to 95% of the pipe diameter may be unstable. This instability can be manifested in a loss of design flow rate and large variations in flow rate. It is also desirable to have an air space above the wastewater for ventilation of gases, such as hydrogen sulfide, that can form. Raw wastewater in sewers has significant amounts

Table 12-3. Froude number, flow characteristics, and depth for the open channel flow regimes.

Flow	Fr	Flow characteristic	Depth
Subcritical	<1	Low inertial flow	$D > D_c$
Critical	$=1$	Inertial force offset by gravitational force	$D = D_c$
Supercritical	>1	High inertial flow	$D < D_c$

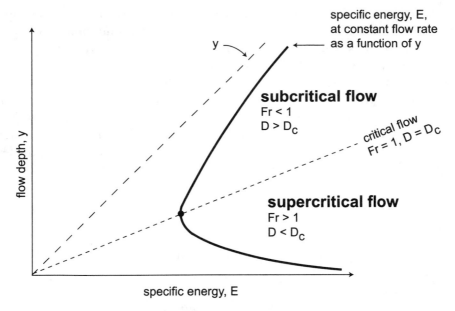

Figure 12-5. Regions of subcritical and supercritical flow on the specific energy plot for flow in an open channel.

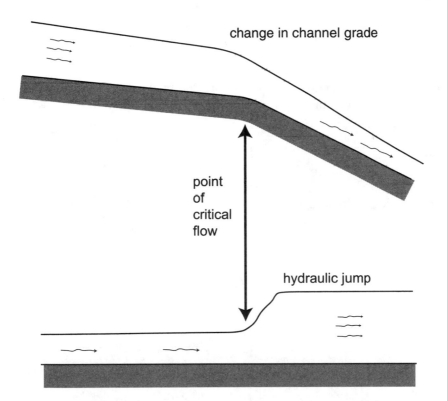

Figure 12-6. Examples of large changes in flow depth with conditions near critical flow in an open channel. *Source:* Adapted from Mays (2000).

of solids that can settle from the flowing stream unless the velocity is high enough to keep them suspended. Sewers should be designed for a minimum of 0.6 m/s (2 ft/s) to keep solids that are organic in nature suspended. To resuspend settled solids, a velocity of 0.8 to 1.1 m/s (approximately 2.5 to 3.5 ft/s) may "flush" out those solids. To be effective, this velocity should be applied on a regular daily basis. Sand and grit is kept suspended with a 0.8 m/s (2.5 ft/s) fluid velocity. The cross-sectional area and slope of the channel or pipe should be chosen to keep within these recommended velocities. Usually the channel should be designed for subcritical conditions, with a flow depth 10% to 15% above the critical depth. This is of particular importance when pipes are more than 50% full.

Example

Calculate the critical depth for a rectangular channel 7.5 ft wide with an 80 ft³/s flow rate.

Solution

Critical flow occurs when

$$\mathrm{Fr} = \frac{V_c}{\sqrt{gD_c}} = 1$$

Because velocity is equal to flow rate divided by flow area, and the flow depth y_c is equal to the hydraulic depth D_c, for a rectangular channel

$$V_c = \frac{Q}{T \cdot y_c}$$

Substituting in for V_c gives

$$\frac{\frac{Q}{(T \cdot y_c)}}{\sqrt{gy_c}} = 1$$

Rearranging leads to

$$y_c = \left(\frac{Q}{T\sqrt{g}} \right)^{2/3} = \left(\frac{80 \ \mathrm{ft^3/s}}{7.5 \ \mathrm{ft} \cdot \sqrt{32.2 \ \mathrm{ft/s^2}}} \right)^{2/3} = 1.5 \ \mathrm{ft}$$

Hydraulic Grade Lines

The total energy at any point in a flowing system may be quantified with Eq. 12-19:

$$E = \frac{P}{\rho g} + z + \frac{V^2}{2g} \tag{12-19}$$

As a fluid flows through the channels, tanks, and other processes that make up a treatment system, energy is lost from friction. So the total energy that the fluid has at the head of the treatment system is greater than the energy at the end. Frictional losses may be calculated with the procedures presented in Chapter 5 for closed-conduit flow and in this chapter for open channel flow, and therefore the energy at each point can be found. This energy grade line (EGL) represents the total energy: the sum of the pressure head, static head, and velocity head as shown in Eq. 12-19 (See Fig. 12-2). Subtracting the velocity head energy from the total energy gives the hydraulic grade line (HGL). The HGL may be depicted on an elevation view of the system (see Fig. 12-2 and Fig. 12-7), showing the height that the fluid will rise along the system, a particularly important issue to prevent flooding in the system. The HGL is generally above the top of the conduit in pressurized closed-conduit flow and below the top of the channel in open channel flow. The distance from the HGL to the top of the channel in open channel flow is termed the *freeboard*. A minimum freeboard of approximately 25 cm (~12 in.) (AWWA and ASCE 1990) is usually considered acceptable with nonturbulent flow that is not near the critical point. A greater freeboard should be called for in channels and tanks if splashing or turbulence is expected.

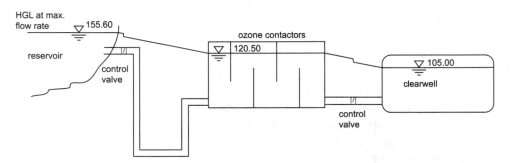

Figure 12-7. Example of a typical hydraulic grade line depicted on a simple water treatment system elevation view.

Symbol List

A	cross-sectional area
C	Chezy coefficient
C_f	flow frictional coefficient
D	hydraulic depth
E	energy
F_f	friction force
F_g	gravity force
Fr	Froude number
h_L	head loss
L	length of control volume
n	Manning's roughness coefficient
P	wetted perimeter; pressure
Q	volumetric fluid flow rate
R	hydraulic radius
S_0	channel bottom slope
S_f	friction head loss, h_L, per unit channel length
T	channel top width (at water surface)
V	fluid velocity
y	liquid depth
γ	specific weight of fluid
ρ	fluid density
θ	angle of inclination of channel bottom
τ_0	shear stress at wall

Problems

1. Wastewater is flowing at a velocity of 1 m/s through a 1.75-m-wide open channel with a rectangular cross section. The walls of the channel are covered with biological slime. The channel bottom slope is 0.0008 m/m. What is the flow depth?

2. A 0.8-m-diameter cast iron pipe (of circular cross section) is flowing with water half-full on a slope of 0.001 m/m. What is the water flow rate?

3. Water flows at 6,800 gpm in a slime-coated triangular-shaped channel (90° included angle) on a slope of 0.008 ft/ft. What is the water depth in the channel? Is the flow subcritical or supercritical?

4. If the flow in Problem 3 is supercritical, what could be done (keeping the same flow rate and slope) to keep the flow subcritical? Be specific.

5. Water must flow at 27,000 gpm through a finished concrete, open rectangular channel on a 0.002 slope. What channel width will maintain a flow depth 15% greater than the critical depth?

References

Avallone, E. A., and Baumeister, T., III, eds. (1996). *Marks' Standard Handbook for Mechanical Engineers*, McGraw-Hill, New York.

AWWA and ASCE (1990). *Water Treatment Plant Design*, McGraw-Hill, New York.

Chow, V. T. (1959). *Open Channel Hydraulics*, McGraw-Hill, New York.

Mays, L. W. (2000). *Water Resources Engineering*, Wiley, Hoboken, NJ.

Properties of Water

Properties of water—SI units

Temperature [C]	Density [kg/m³]	Internal energy [kJ/kg]	Enthalpy [kJ/kg]	Speed of sound [m/s]	Vapor pressure [kPa]	Viscosity [cP]
0.01	999.84	0.0019	0.10	1402.4	0.6117	1.791
2	999.94	8.39	8.49	1412.2	0.7060	1.673
4	999.97	16.81	16.91	1421.6	0.8136	1.567
6	999.94	25.22	25.32	1430.6	0.9354	1.471
8	999.85	33.62	33.73	1439.1	1.0730	1.385
10	999.70	42.02	42.12	1447.3	1.2282	1.306
12	999.50	50.41	50.51	1455.0	1.4028	1.234
14	999.25	58.79	58.89	1462.4	1.5990	1.168
16	998.95	67.16	67.27	1469.4	1.8188	1.108
18	998.60	75.54	75.64	1476.0	2.0647	1.053
20	998.21	83.91	84.01	1482.3	2.3393	1.002
22	997.77	92.27	92.37	1488.3	2.6453	0.954
24	997.30	100.64	100.74	1494.0	2.9858	0.911
26	996.79	109.00	109.10	1499.3	3.3639	0.870
28	996.24	117.36	117.46	1504.4	3.7831	0.832
30	995.65	125.72	125.82	1509.2	4.2470	0.797
32	995.03	134.08	134.18	1513.6	4.7596	0.765
34	994.37	142.44	142.54	1517.8	5.3251	0.734
36	993.69	150.80	150.90	1521.8	5.9479	0.705
38	992.97	159.16	159.26	1525.5	6.6328	0.678
40	992.22	167.51	167.62	1528.9	7.3849	0.653
42	991.44	175.87	175.98	1532.1	8.2096	0.629
44	990.63	184.23	184.33	1535.1	9.1124	0.607
46	989.79	192.59	192.70	1537.8	10.099	0.586
48	988.93	200.95	201.06	1540.3	11.177	0.566
50	988.04	209.32	209.42	1542.6	12.352	0.547

52	987.12	217.68	217.78	1544.7	13.631	0.529
54	986.17	226.04	226.15	1546.5	15.022	0.512
56	985.21	234.41	234.51	1548.2	16.533	0.496
58	984.21	242.78	242.88	1549.7	18.171	0.481
60	983.20	251.15	251.25	1551.0	19.946	0.466
62	982.16	259.52	259.62	1552.1	21.867	0.453
64	981.09	267.89	267.99	1553.0	23.943	0.440
66	980.00	276.26	276.37	1553.8	26.183	0.427
68	978.90	284.64	284.74	1554.3	28.599	0.415
70	977.76	293.02	293.12	1554.7	31.201	0.404
72	976.61	301.40	301.50	1555.0	34.000	0.393
74	975.44	309.78	309.89	1555.1	37.009	0.383
76	974.24	318.17	318.27	1555.0	40.239	0.373
78	973.03	326.56	326.66	1554.8	43.703	0.363
80	971.79	334.95	335.06	1554.4	47.414	0.354
82	970.53	343.35	343.45	1553.9	51.387	0.346
84	969.26	351.74	351.85	1553.3	55.635	0.337
86	967.96	360.15	360.25	1552.5	60.173	0.329
88	966.64	368.55	368.65	1551.5	65.017	0.322
90	965.31	376.96	377.06	1550.5	70.182	0.314
92	963.96	385.37	385.48	1549.2	75.684	0.307
94	962.58	393.79	393.89	1547.9	81.541	0.301
96	961.19	402.21	402.31	1546.5	87.771	0.294
98	959.78	410.63	410.74	1544.9	94.390	0.288
99.974	958.37	418.95	419.06	1543.2	101.320	0.282

Data from: E. W. Lemmon, M. O. McLinden, and D. G. Friend, "Thermophysical Properties of Fluid Systems," in *NIST Chemistry WebBook*, NIST Standard Reference Database Number 69, edited by P. J. Linstrom and W. G. Mallard, June 2005, National Institute of Standards and Technology, Gaithersburg, MD (http://webbook.nist.gov).

Properties of water—US customary units

Temperature [F]	Density [lbm/ft³]	Internal energy [BTU/lbm]	Enthalpy [BTU/lbm]	Speed of sound [ft/s]	Vapor pressure [atm]	Viscosity [cP]
2	999.94	8.39	8.49	1412.2	0.7060	1.673
32	62.418	0.001	0.044	4601.2	0.00604	1.791
35	62.424	3.006	3.050	4628.0	0.00680	1.692
40	62.426	8.037	8.081	4670.8	0.00828	1.545
45	62.421	13.060	13.104	4710.8	0.01004	1.417
50	62.409	18.076	18.120	4748.3	0.01212	1.306
55	62.391	23.087	23.131	4783.2	0.01457	1.208
60	62.367	28.094	28.137	4815.8	0.01745	1.121
65	62.337	33.097	33.140	4846.2	0.02081	1.044
70	62.301	38.097	38.141	4874.4	0.02473	0.975
75	62.261	43.095	43.139	4900.5	0.02927	0.913
80	62.216	48.092	48.136	4924.7	0.03453	0.857
85	62.167	53.087	53.131	4947.0	0.04060	0.807
90	62.113	58.082	58.126	4967.6	0.04757	0.761
95	62.055	63.077	63.121	4986.4	0.05555	0.719
100	61.994	68.071	68.115	5003.5	0.06468	0.681
105	61.929	73.065	73.109	5019.1	0.07507	0.646
110	61.860	78.060	78.104	5033.1	0.08687	0.614
115	61.788	83.055	83.099	5045.7	0.10024	0.585
120	61.712	88.051	88.096	5056.9	0.11534	0.557
125	61.633	93.048	93.092	5066.7	0.13235	0.532

130	61.552	98.046	98.090	5075.2	0.15147	0.508
135	61.467	103.050	103.090	5082.4	0.17289	0.487
140	61.379	108.050	108.090	5088.5	0.19686	0.466
145	61.288	113.050	113.090	5093.4	0.22359	0.447
150	61.195	118.050	118.100	5097.1	0.25335	0.430
155	61.099	123.060	123.100	5099.8	0.28640	0.413
160	61.000	128.060	128.110	5101.4	0.32303	0.398
165	60.899	133.070	133.120	5102.0	0.36354	0.383
170	60.795	138.080	138.130	5101.6	0.40826	0.370
175	60.688	143.100	143.140	5100.2	0.45752	0.357
180	60.580	148.110	148.160	5097.9	0.51167	0.345
185	60.468	153.130	153.180	5094.7	0.57110	0.333
190	60.355	158.150	158.200	5090.7	0.63620	0.323
195	60.239	163.180	163.220	5085.7	0.70738	0.312
200	60.121	168.200	168.250	5080.0	0.78508	0.303
205	60.000	173.240	173.280	5073.4	0.86976	0.294
210	59.878	178.270	178.320	5066.0	0.96187	0.285
212	59.829	180.240	180.280	5062.9	1.00090	0.282

Data from: E. W. Lemmon, M. O. McLinden and D. G. Friend, "Thermophysical Properties of Fluid Systems," in *NIST Chemistry WebBook*, NIST Standard Reference Database Number 69, edited by P. J. Linstrom and W. G. Mallard, June 2005, National Institute of Standards and Technology, Gaithersburg, MD (http://webbook.nist.gov).

Index